The Laser in America, 1950–1970

The Laser in America, 1950–1970

Joan Lisa Bromberg

The MIT Press
Cambridge, Massachusetts
London, England

This book is an outgrowth of the Laser History Project, which was sponsored by the American Physical Society, the IEEE Lasers and Electro-Optics Society, the Laser Institute of America, and the Optical Society of America. It is not, however, an official document of these societies or of any of the public or private organizations that have contributed financial support to the project. The opinions and interpretations in the book are those of the authors.

This book was typeset in New Baskerville and printed and bound in the United States of America

Library of Congress Cataloging-in-Publication Data

Bromberg, Joan Lisa.
    The Laser in America, 1950–1970 / Joan Lisa Bromberg.
      p. cm.
    Includes bibliographical references and index.
    ISBN 978-0-262-02318-4 (hc.:alk.paper) 978-0-262-51980-9 (pb)

    1. Lasers—History—20th century.   I. Title.
TA1677.B76   19921990
621.36'.6—dc20
      90-44204   CIP

The MIT Press is pleased to keep this title available in print by manufacturing single copies, on demand, via digital printing technology.

Contents

Epilogue: The Laser Now and in the Future   228
A. H. Guenther, H. R. Kressel, and W. F. Krupke

Foreword

Arthur H. Guenther
for the Laser History Project Advisory Committee

This volume is an outgrowth of the Laser History Project, which was initiated in 1982, at the suggestion of the Laser Institute of America, by four U.S. scientific societies: the American Physical Society, the Institute of Electrical and Electronics Engineers' Quantum Electronics and Applications Society (now the Lasers and Electro-Optics Society), the Laser Institute of America, and the Optical Society of America. The rationale was the approaching 25th anniversary of the first operating lasers. The laser was rapidly becoming a standard tool in science and in military, industrial, and consumer technology. Yet most of the men and women who had done the pioneering research were still alive and vigorous. The societies recognized that this was an opportune time to collect both oral histories and documentation that would be indispensable for future historical treatments. A secondary objective of the project was to collate its findings with other resource materials, to yield an external perspective on how science and society dealt with the rapid, almost explosive, development and application of the laser. It was hoped that this would become a good object lesson for society's handling of future discoveries.

Between 1982 and 1988 the project conducted more than 80 interviews, mostly with U.S. pioneers, but also with Soviet Nobel laureates N. G. Basov and A. M. Prokhorov. It collected bibliographies, vitas, memoirs, and unpublished documents from about 150 persons who had worked in the field. It identified repositories of relevant documents and artifacts as well. It also sponsored a symposium on the 25th anniversary of the 1960 lasers, and it suggested, and then served as consultant for, a major Smithsonian Institution traveling exhibit on the history of lasers, funded by the Optical Society of America and the IEEE Lasers and Electro-Optics Society. The exhibit circulated in the United States for two years.

The work of the Laser History Project has been overseen by an Advisory Committee consisting of myself as chair, Nicolaas Bloembergen, John A. Armstrong, and Joseph A. Giordmaine for the American Physical Society; William B. Bridges, James P. Gordon, and Henry Kressel for the IEEE Lasers and Electro-Optics Society; C. Breck Hitz, John F. Ready, and R. James Rockwell jr. for the Laser Institute of America; and William Condell, Alexander J. Glass, and Emil Wolf for the Optical Society. Joan Warnow-Blewett and Spencer R. Weart, directors of the Center for History of Physics at the American Institute of Physics also served on the committee, as did the director of the IEEE Center for the History of Electrical Engineering and Bernard S. Finn, curator of electricity at the Smithsonian Institution's National Museum of American History. Helen W. Samuels, archivist for the Massachusetts Institute of Technology, and Robert W. Seidel, director of the Bradbury Museum at the Los Alamos National Laboratory, rounded out its membership. Haynes A. Lee jr. managed the project's finances. Cooperation with the AIP and IEEE centers and with the Smithsonian Institution has extended beyond mere representation on the Advisory Committee. The AIP Center has transcribed the bulk of the interviews as a contribution-in-kind; it has also supervised the microfilming of the project's collection, and is publishing a catalog describing, but extending beyond, our collection. The Smithsonian Institution celebrated the project's completion in late 1988 with a special luncheon and exhibit. The IEEE and AIP Centers have canvased leading laser scientists and laser research organizations to alert them to the importance of saving unpublished materials.

The mixed composition of the Advisory Committee led to a great appreciation on the part of both scientists and historians for each other's concerns and perspectives; the constant attention and input of both groups was essential for the project to meet its goal. Discussions were always frank between the members and between the project director and myself. On behalf of the Advisory Committee, I would like to compliment Joan Bromberg for her thoroughness, dedication, and professionalism in this, at times, contentious undertaking.

The project had a separate fund-raising committee, the Laser History Council, composed of John H. Marburger III as chair, Nicolaas Bloembergen, Herbert M. Dwight, Arthur Kantrowitz, Arthur L. Schawlow, George F. Smith, and Charles H. Townes. The funding for the project came from a variety of private and public organizations. The four founding societies put up the seed money and made subsequent contributions. Thirteen private corporations gave donations: AVCO Corporation, Aerospace Corporation, Coherent, Inc., Cooper LaserSonics, Inc., General Electric Company, General Motors Corporation, Hughes Aircraft Company, International Business Machines, KMS Fusion, Inc., Newport Corporation, Raytheon, United Technologies Research Center, and XMR, Inc. The project received grants from The U.S. Air Force Office of Scientific Research, the U.S. Army Research Office, the U.S. Office of Naval Research, and the U.S. Department of Energy's Office of Inertial Fusion. The National Science Foundation, the National Endowment for the Humanities, and the Alfred P. Sloan Foundation all awarded grants specifically earmarked for historical research and writing.

Historical writing has played an important role in the project from the start. The Advisory Committee saw this activity as being potentially useful to both scientists and historians. We have sought to give scientists accounts to draw upon for class lectures, textbooks, and public presentations, and historians a framework within which to tackle specific issues that might have implications not only for the evolution of science and technology, but for business history, science policy, and the sociology of science. Over the years, project director Joan Lisa Bromberg and Robert W. Seidel (who served for two years as the Project's Research Historian) have published articles deriving from the project in scientific and historical periodicals. The present book is the culmination of this phase of project activity.

It is important to emphasize that while *The Laser in America* was written as part of the project, and while members of the Advisory Committee and the Laser History Council read portions of the manuscript and discussed them with the authors, our policy has been to allow the authors complete freedom to offer their own interpretations. Many differences of opinion remain, and the book should not be construed as representing the opinions of the project's Advisory Committee or Council, sponsoring societies, or contributors. Nevertheless, the Advisory Committee views the appearance of this book with considerable satisfaction. For those of us who are laser scientists, much of it rings true to the spirit of what we experienced. We believe that it will give the reader an insight into how it feels to participate in the early, exhilarating phases of a new branch of science.

Preface

The family of devices that comprises masers and lasers is only 35 years old. It is not surprising, therefore, that most previous accounts of their history have been written by scientists and journalists and focus on technical aspects of the story. This book will offer the viewpoint of a historian and will therefore attempt to place the technical developments in their social context.

I owe a debt to those scientists who have written on the laser and to the large number of historians who have discussed the environment for science and technology in post–World War II America. My greatest intellectual debt, however, is to Paul Forman. I have made use of his articles and exhibits on atomic clocks, a technology closely related to masers.[1] I have incorporated into the theoretical framework of this book the ideas of his article "Beyond Quantum Mechanics: National Security as a Basis for Physical Research in the United States, 1940–1969."[2] I have profited from his counsel. And I have used his writing to keep before me the vision of how a master historian practices the craft.

The book draws on a number of sources. First are published works, including scientific papers and books as well as articles in the trade and business press. Then there are the taped interviews I conducted with 50 maser and laser pioneers and the memoirs I collected from about 100 others. These tapes and memoirs, along with related materials gathered under the auspices of the Laser History Project, are available to scholars in the Sources for the History of Lasers (SHL) at the Niels Bohr Library of the American Institute of Physics (AIP) in New York City. A microfilm edition will be available to other repositories through the AIP's Center for the History of Physics. The Center, under the direction of Joan Warnow-Blewett, is publishing a catalog of sources for laser history that will include the SHL materials.

I have also consulted several manuscript collections. These include the papers of Charles H. Townes, which will eventually be deposited in the collections of the Bancroft Library of the University of California at Berkeley; the Rudolf Kompfner papers at the AT&T Archives (a unit of the AT&T Library Network) in Warren, New Jersey; the papers of the Columbia University Physics Department at the Columbia University Archives; and papers at the Hughes Aircraft Company Archives.

Paul Forman kindly made available to me reproductions he had obtained from the Signal Corps Records at the Federal Records Center in St. Louis and from the papers of Harold Lyons. Many scientists also generously opened their files to me or sent me copies of documents from their personal papers. Lawyers for a number of the litigants in patent cases supplied me with copies of depositions or briefs. These items and other sources are all indicated in the notes.

For information on classified military research, I relied heavily on two important papers by Robert W. Seidel, the research historian for the Laser History Project: "From Glow to Flow: A History of Military Laser Research and Development" and "How the Military Responded to the Laser."[3] These papers were based on research in classified archives and on 25 classified interviews. Some of these interviews have since been declassified and are included in the SHL. Others are in the process of declassification, and the remainder are on deposit at the Air Force Weapons Laboratory Historian's Office at Kirtland Air Force Base, New Mexico, where they are available to people with the necessary clearance.

The character of my sources, taken collectively, required that I confine myself almost entirely to work done in the United States. There is a second reason for this focus: I have chosen to write a book that is as much a study of the U.S. scientific establishment, as reflected in the light of laser research, as a story of maser and laser science and technology. This does not mean that I judged foreign contributions to be negligible. On the contrary, even the historian who looks at American work alone sees continually the impact of advances made in Europe and the Soviet Union in the 1950s and 1960s. Clearly we need histories of maser and laser research in other national contexts. Then we may compare the effect of varying environments on research directions and attack a range of supranational questions such as the diffusion of scientific information between nations and within nations.

These comments on sources will also give a partial idea of the extent of the obligations I have incurred in writing this book. I owe a great debt to the many people who set aside time for interviews and who allowed me access to their files or themselves hunted out documents for my use. I am also in debt to the people who read and commented on my discussions of their work or the work of colleagues. And I am grateful to my own colleagues in the history of science and technology for a variety of suggestions of primary and secondary sources.

I appreciate the work of the archivists who helped me through a number of collections, including Bruce W. Henry of the Hughes Aircraft Company, Anita Newell and John Coltman of the Westinghouse R&D Center, Marcy Goldstein, Norma McCormick, and Jeffrey L. Sturchio of the AT&T Bell Laboratories Archives, Mary Murphy of the MIT Lincoln Laboratory Library Archives, and Anne Milbrooke of the United Technologies Archive. For providing me with legal briefs and depositions, I thank Richard I. Samuel of Patlex Corporation, Samuel Dworetsky of AT&T, Dana M. Raymond of Brumbaugh, Graves, Donahue and Raymond, and Tim G. Jagodzinski of General Motors Corporation. I appreciate the courtesy of Jeff Hecht, Richard Cunningham, Michael Wolff, and James Cavuoto in making available unedited transcripts of their interviews with laser pioneers. Finn Aaserud and Virginia French of the Center for the History of Physics oversaw the processing of the project's interviews and aided me in innumerable ways.

This book could not have been written if a number of institutions in the Boston area had not granted me access to their libraries. I am grateful to the Science, Technology, and Society Program, the Electrical Engineering and Computer Science Department, and the George Harrison Laboratory of Spectroscopy of the Massachusetts Institute of Technology for appointing me a visiting scholar, and to the History of Science Department of Harvard University for making me an associate. I also thank the director and staff of the excellent Air Force Geophysical Laboratory Research Library at Hanscom Air Force Base, as well as Randall W. Bergmann, manager of the Defense Technical Information Center at Hanscom.

My work on the book was supported by grants from the National Science Foundation, the National Endowment for the Humanities, and the Alfred P. Sloan Foundation, as well as by the project's sponsoring societies—the American Physical Society, the IEEE Lasers and Electro-Optics Society, the Laser Institute of America, and the Optical Society of America—and the contributors listed in the foreword.

There are two people to whom I owe a special debt. Janet L. French entered the manuscript on the computer over five years and brought me chicken soup when I was sick and pumpkin bread when I was well. My editor, Larry Cohen, guided the book along with care.

Many scientists and historians read one or another draft and made comments and corrections of great value. I should particularly like to mention William B. Bridges, Paul Forman, Elsa M. Garmire, Bruce W. Henry, Daniel J. Kevles, John F. Ready, Arthur L. Schawlow, Robert W. Seidel, George F. Smith, Spencer R. Weart, and Emil Wolf. The mistakes and misinterpretations that remain are solely my responsibility.

Finally, I am profoundly grateful to Arthur H. Guenther and the members of the advisory committee and to Haynes A. Lee jr., the project's financial manager. Their friendship and support gave me encouragement, and their astute maneuvering through the shoals that lie in wait for any project like this one made it possible to get the work finished.

The Laser in America, 1950–1970

1

Introduction

The story I shall tell in these pages has the traditional structure of a biography. It begins with the hero's progenitor, in this case the maser. It goes on to discuss conception, birth, the vigorous growth of the child, and, finally, the ripening of the adolescent toward maturity. It was a fortunate childhood, for the laser was born in a time and place that were especially favorable for the development of such an electronic device.

Electronics research benefited from two trends that characterized the United States in the period after 1950. First, resources for research and development in general were rising dramatically. Total annual expenditures on R&D rose from $2.6 billion in 1949 to $6.2 billion in 1955 and $12.4 billion in 1959,[1] while the proportion of gross national product expended on R&D rose from 1% in 1950 to 2.8% in 1960.[2] Second, the market for electronics was growing rapidly, with total sales increasing from $3.4 billion in 1950 to $11.4 billion in 1960, in constant 1967 dollars.[3]

Many factors spurred interest in electronics R&D. One was the example of the transistor, which clearly demonstrated the payoffs that awaited the right product. Another was the growth of industrial applications of electronics. A third was the unusually high prestige that science and technology were enjoying. And a fourth factor, one that gave support and direction to electronics research throughout the period, was federal military spending.

The period I address in this book, 1950–1970, began with the Korean War, which marked a nodal point in military spending. The Truman administration used the Korean crisis of June 1950 to increase the military budget from about $13 billion per year to $50 billion per year. Dwight David Eisenhower, who took office in 1953, trimmed the budget somewhat, but it still remained at two to three times pre-Korea levels.[4]

Military expenditures spurred the entire economy, but they acted preferentially to accelerate electronics. One reason for this was that aircraft were the weapons of choice in the early 1950s. Truman and Eisenhower both saw atomic bombs, delivered by Air Force bombers, as the most effective way to maintain U.S. military power without straining the economy.[5] Already by 1950, electronics comprised one-third to one-half of the cost of a military airplane.[6] When, starting about 1955, the missile revolution took hold, electronics increased in importance, since guided missiles needed electronics even more than aircraft. Thus, while consumer electronics sales held fairly constant between 1950 and 1960, from $1.9 billion to $2.1 billion in constant 1967 dollars, and industrial sales increased nearly fivefold, from $440 million to $2.1 billion, government electronics sales, which were largely military, increased nearly eightfold, from $830 million to $6.5 billion. It was military spending that made electronics one of the most rapidly growing sectors of the economy in the 1950s.[7]

At the same time, the U.S. military establishment had committed itself more broadly to the notion of a permanent technological revolution in weaponry and was pouring money into research and development. Department of Defense (DoD) expenditures for R&D itself and for R&D infrastructure rose from $590 million in fiscal year 1948 to $2.6 billion in FY 1956. (In the same period, federal R&D expenditures outside the DoD and the Atomic Energy Commission went from $150 million to $330 million.)[8] Military electronics R&D funding rose correspondingly.

Research on masers began in early 1951, shortly after the start of the Korean War, and DoD money was vital to their creation and development during the 1950s. The Army Signal Corps, the Air Force, and the Office of Naval Research supported the construction of the first maser at Columbia University through the Joint Services Electronics Program,[9] and they individually supported maser R&D contracts at other universities and at industrial laboratories.[10]

Given the availability of research funds, the number of industrial laboratories in the United States grew from about 3,000 in 1950 to about 5,400 in 1960.[11] The climate made it easy for consulting companies to form and to sustain themselves. Venture capital was readily available for high-technology firms, in part precisely because investors anticipated that the companies they financed would have military markets.[12] One example of a firm that had its origins in the climate created by DoD spending was Technical Research Group, which was to play an important role in the history of the laser in 1958 and after.

Many established companies were also building new central laboratories in this period, often with campuslike amenities. Other large companies reorganized existing laboratories to place more emphasis on electronics. Airframe and air engine companies set up electronics laboratories in order to enter the new field of missiles.[13] Optics companies reoriented their research to focus on the emerging area of electro-optics. The military market was one spur. The need to compete in industrial markets in an era of accelerating technological innovation was another. Still a third factor was ideological: This was a period in which economists viewed R&D as the mainspring for economic growth,[14] while cold warriors saw U.S. preeminence in science and technology as an essential weapon in the struggle against the Soviet Union.

For new laboratories, the maser and later the laser were attractive research areas. Electronics was central to their civilian and military markets, and maser and laser research, positioned at the scientific forefront of electronics, was a field in which reputations could be made for both a laboratory and its personnel. IBM's T.J. Watson Research Center, which was to contribute important work on solid-state and semiconductor lasers, was established in this period. So was the Avco Everett Research Laboratory, birthplace of the gas dynamic laser. Hughes Aircraft Company's Research Laboratories, United Aircraft Research Laboratories, and the research laboratories of the American Optical Company were all expanded in the 1950s in directions that would allow them to become centers for laser research.

This was also a period in which the number of scientists was increasing, while their focus was shifting from universities to industry. The number of research scientists and engineers rose from 144,000 in 1949 to 223,000 in 1954. The number of physicists increased by more than a third.[15] Nevertheless, industrial laboratory space was expanding so quickly that there was a shortage of trained technical personnel throughout the decade. The shortage of people in relation to the increasing number of technical jobs combined with the public prestige of science to force employers to provide ever more favorable working conditions. This included considerable freedom in the choice of research topics.[16] Industrial scientists were often allowed to devote much of their energy to attacking problems posed by their scientific disciplines, and they garnered their rewards from within their professions as well as from their companies. This, too, was a situation that would be advantageous for the development of masers and lasers, since these devices were intriguing to scientists but often promised profits only in the long run.

The economic and social environment, already favorable, grew even more hospitable to electronics R&D after October 1957, when the Soviet Union orbited its first Sputnik satellite. The rate of growth of the total U.S. R&D budget, which had been under 15% a year in 1958, spurted to over 40% in 1959. The rise in funding for federal organizations was often precipitous. For example, the Air Force Office of Scientific Research's fiscal year 1958 appropriation was $16.3 million; for fiscal year 1959, it was $27 million.[17] Additional federal agencies with R&D missions were created. The National Aeronautics and Space Agency (NASA) was one. Another was the Advanced Research Projects Agency (ARPA), established by the new secretary of defense, Neil McElroy, in order to sidestep interservice rivalry over space projects. ARPA metamorphosed, by 1959, into an agency for the support of basic research into advanced military technology.[18]

Sputnik, though a civilian satellite, underlined the Soviet Union's ability to loft missiles and military satellites. For this reason, research on measures to detect and destroy ballistic missiles and satellites now became a central interest for DoD research agencies. The maser profited, for it held promise as a component in long-range radars and other detection apparatuses. From the laser's point of view, the timing could hardly have been better, since the research that transformed it from an idea into an operating device began in the summer of 1958. The military wanted to know if laser beam weapons might find a useful place in the armamentarium of antiballistic missiles. It was for this reason, for example, that Technical Research Group's laser program was funded munificently by ARPA, with a budget three times higher than the company had solicited.

Federal largesse for military science and technology continued under the administration of John F. Kennedy. Kennedy was elected in 1960, the year the first lasers were made to operate. When Kennedy and his secretary of defense, Robert McNamara, took office in 1961, one of their first steps was to accelerate the procurement of strategic systems. The goal was to offset a supposed "missile gap" with the Soviets that Kennedy had emphasized in his campaign rhetoric. Shortly into his term, Kennedy also asked Congress to fund the Apollo program to place a U.S. astronaut on the moon.

Both the stepped-up missile program and the space program were boons to the electronics industry. For the laser in particular, they meant higher levels of electronics R&D funding at the very time when proof had been provided that lasers could indeed be made to work. This situation encouraged the start-up of laser firms and of laser divisions within established companies. R&D contracts became an important component of the funding of these businesses. The commercial laser industry in turn fostered laser research, for it provided off-the-shelf equipment to the scientists who were pursuing laser science or research into laser applications. The industry also made technological contributions of its own. These were above all contributions to the reliability and durability of lasers, but commercial groups were also responsible for some of the wavelengths, laser types, and novel means of controlling the laser beam that were discovered during the 1960s.

At the end of 1961, an American Optical Company scientist invented the glass laser, a variant that promised higher power than any type operated to that date. This technological achievement helped convince the Department of Defense to ramp up its programs in high-energy laser research. Thus, technical advance and military interest continued to feed off each other in the early Kennedy years.

In 1963–1965, in the last year of the Kennedy administration and the first years of Lyndon B. Johnson's administration, overall military strategy changed. Secretary McNamara continued the movement, begun under Eisenhower, away from a policy of massive nuclear retaliation and toward a repudiation of any first nuclear strike. He chose the option of a second-strike strategic nuclear force, large enough to deter Soviet nuclear attack but not necessarily superior in size to Soviet forces. The partial test-ban treaty signed in 1963 and the ensuing détente reinforced this strategy. At the same time, the administration moved to build up conventional forces, which had been neglected under Eisenhower, after Kennedy and his advisers identified wars within the third world as the most likely arena of the Soviet-American struggle.[19] Defense spending therefore began to fall off in the early years of the Johnson administration. Although spending on NASA space programs took up the slack, an absolute decline was widely forecast, and electronics firms therefore scurried to shift into civilian markets. This tactic appeared the more sensible in that there was a long-term trend, reaching back into the 1950s, of expansion in the industrial electronics market.[20]

The cutback in defense electronics was short-lived. It was reversed around 1966, as the Vietnam war expanded. This war, however, provided a qualitatively different defense market. Procurement and R&D contracts for strategic missiles remained low. The space program was also cut back. The electronics money now went for systems such as ground-to-ground and ground-to-air communications, night-vision apparatus that could locate enemy emplacements through jungle cover, and range-finding and target-locating devices for artillery, small planes, and tanks.[21] Laser R&D money was shifted into the development of range finders, smart bombs, and other tactical aids. These products provided the first "mass market" for laser systems; purchases by the armed services were in the hundreds of systems as the decade came to an end.

Even as these weapons were arriving at battlefields, however, a golden age for U.S. science and technology was drawing to a close. After 1968, total U.S. spending on R&D (in constant dollars) began to decline. Federal spending decreased even more than nonfederal expenditures; the responsible factors were the winding down of the Vietnam war, the Nixon administration's pursuit of détente, the successful completion of the Apollo program, and government austerity measures in the face of the economic downturn of the early 1970s. Industry also cut back on its overall funding and began to abandon basic research in favor of short-term development for immediate goals.[22]

Accompanying these changes was a shift in public mood. From 1968 on, opposition to the war grew. The general public rejected the war as inconclusive and entailing too large a sacrifice, while academics and other elites opposed it on moral as well as practical grounds.[23] As a result, military R&D became ethically questionable. The civil-rights movement of the mid-1960s had already focused attention on the needs of the cities and spawned a movement to direct national resources toward civilian rather than military goals. Growing awareness, first, of the fragility of the environment and, somewhat later, of the exhaustibility of energy sources reinforced such demands. Government R&D spending began a shift that was to last through the 1970s, away from military work and toward energy, environment, and health. About 85% of federal R&D money went for defense and space in the mid-1960s, as against only about 65% in 1979.[24]

These events were all refracted into the laser world. As industry shifted its R&D emphasis from long- to short-term programs, industrial laser scientists found themselves increasingly restricted in their freedom to follow their own hunches about profitable research directions. Laser sales to the military continued to increase, but they grew more slowly than civilian sales. The trade magazine *Laser Focus* estimated that the military share went from 63% in 1969 to 58% in 1970 and 55% in 1971.[25] This circumstance, and the new ideology against weapons research and for socially useful research, had an effect. The laser community responded to the swords-into-plowshares mood of the early 1970s with a flurry of R&D on laser pollution-monitoring devices and a pair of responses to the energy crisis: laser fusion energy and laser separation of uranium isotopes to create enriched fuel for nuclear reactors. The decrease in federal R&D expenditures also had an indirect effect: As funds for university research declined, purchases of lasers as research instruments fell drastically. As a result, the laser industry began to change its marketing strategy, from selling stand-alone lasers to marketing complete laser systems for industrial applications. By 1970, the point at which this narrative leaves off, the enchanted childhood was over. The U.S. laser enterprise had now to function in a more severe climate.

The fact that a broad variety of institutions participated in maser and laser research is already implicit in the foregoing discussion. Defense contractors, such as Hughes Aircraft Company and United Aircraft, and firms with a mix of commercial and military products, such as IBM and American Optical Company, were involved. So were new firms founded by scientist-entrepreneurs. Government laboratories mounted programs. Maser and laser groups formed within academic departments of engineering, physics, biology, and chemistry. University-run, government-funded laboratories such as the Columbia Radiation Laboratory, the University of Michigan's Willow Run Laboratories, MIT's Lincoln Laboratory, and the California Institute of Technology's Jet Propulsion Laboratory joined in.

Each institutional type has its own social environment, and in these divergent soils, different flowers could bloom. A salient feature of university research is that the labor is done by graduate students. On the one hand, this offers faculty a supply of cheap and expert labor, capable of nursing along finicky, unreliable equipment. On the other hand, it limits them to projects that will yield solid, scientifically respectable theses. In particular, if they choose to develop a new instrument, it must be an instrument on which new science can be done. It is also incumbent upon professors directing research projects not to undertake work in areas so highly competitive that their students are in danger of being scooped before they are finished. It is not surprising that the Columbia Radiation Laboratory was the site for the construction of the first maser at a time, 1951–1954, when the maser was an idea in only a handful of minds and competition was negligible.

Industrial laboratories characteristically have more ample equipment than universities. A Bell Telephone Laboratories can draw on its own or the government's financial resources; research teams at small firms can sometimes borrow powerful tools from the production line. Moreover, industrial scientists are well positioned to enter competitive areas. Indeed, they are sometimes pressured to do so by management. Not surprisingly, then, it was chiefly industrial teams that participated in the intense rivalry to build the first semiconductor laser in 1962. Maser and laser research benefited from the fact that a variety of research sites were available, because it meant that a wider group of research questions and applications were pursued.[26]

The multiplicity of research sites also confers a benefit upon the historian. Precisely because so many sectors of the U.S. high-technology establishment were involved in their development, the maser and laser become probes with which we can map that establishment. Through specific case studies, we can examine how the various sectors functioned. By focusing on the interactions of groups from different sectors under the concrete circumstances of laser work, we can get hold of the links that bound the sectors together. The historian who undertakes a biography of the laser thus finds herself in a position to paint a portrait of U.S. research in a critical period. This double task is the one I propose to undertake. Indeed, it is precisely the fact that academic scientists and engineers, federal agencies and departments, and industry were all involved, and all interacting with each other as well as with the technical developments, that gives the history of masers and lasers in the United States its larger interest.

In the following chapters I shall tell the story of the maser and laser in the United States and explore the character of the research system within which this technical work unfolded. The first section of chapter 2 describes the scientific and military interest in generators and amplifiers of short-wave radiation around 1950, which formed the context within which the idea of the maser was first advanced. It details the construction of the ammonia beam maser at Columbia University and then discusses the web of interrelations among university, military, and industrial scientists that governed the spread of beam maser research through 1955.

As an amplifier, the ammonia maser was something of a tease; it offered unparalleled sensitivity, but it lacked the width of frequency response and strength of amplification that would make it widely useful. The solution was to translate the maser principle from molecular beams to solids. The second section of chapter 2 traces the work of the Americans who took part in this research. The major achievements were the three-level, continuous, solid-state maser and the demonstration that synthetic ruby made a superior maser material. At this point, the electrical engineers took notice, and this section also deals, in a preliminary way, with the novel perspectives engineers brought to the field.

Solid-state masers were quickly applied to radio and radar astronomy, military radar, and the emerging field of satellite communications. This is the subject of the third section of chapter 2. Among the radio astronomy projects were some that bore on the most fundamental questions of "pure" science. But it was characteristic of this period that these projects were not pursued in isolation from applications-oriented research. Military, commercial, and scientific research projects were often carried out at the same site, with the same equipment, under industrial or military laboratory managers who believed that these diverse studies could fructify each other. Meanwhile, new types of amplifiers were being discovered that were almost as sensitive as, and more convenient than, masers. The enthusiasm that had greeted the maser was rapidly supplanted by a more judicious weighing of its pros and cons. By the early 1960s, masers had come to fill an important but limited

niche among microwave receivers for applications in which exceptional sensitivity was needed and complexity could be tolerated.

Chapter 3 describes the invention of the laser and the construction of its first exemplars. The theoretical work done in the United
States had as its context a widely shared interest in pushing the
maser, which then operated at wavelengths ranging from 1 to 21
centimeters, into the region of millimeter, submillimeter, and far-
infrared wavelengths. Charles H. Townes of Columbia University
was working on this problem in the late summer of 1957 when he
had the insight that it might prove easier to leap over the whole
spectrum from millimeter through midinfrared wavelengths and
go directly to visible and near-infrared wavelengths. The first
section of chapter 3 explores Townes's ideas and his collaboration
with Arthur L. Schawlow of Bell Telephone Laboratories. It com-
pares their work with that of R. Gordon Gould, a Columbia
University graduate student who joined the Long Island company
of TRG, Inc., shortly after he recorded his first laser ideas in
November 1957.

The second section of chapter 3 details the programs to construct
a laser that were initiated in a half dozen U.S. laboratories, largely
inspired by the work of Townes and Schawlow. Each group pursued
a strategy shaped by the experience and scientific style of its leaders
and the orientation of the institution within which it was situated.
Townes at Columbia pursued a potassium vapor laser. Two independent groups at Bell Laboratories worked on noble-gas lasers. A
small group at the American Optical Company worked on glass
lasers. TRG was munificently funded by the Advanced Research
Projects Agency to work on several schemes in parallel. IBM
investigators used rare-earth-doped calcium fluoride crystals. The
first operating laser was achieved in June 1960 by Theodore H.
Maiman of Hughes Research Laboratories. His success was unexpected because the material he used, maser-ruby, was widely believed to be unpromising. And his difficulties in getting published
were to be an embarrassment to the American physics community.

The creation of operating lasers initiated a research boom. A
boom is a prime site for the historian interested in excavating the
structure of a research establishment. From the motivations that
drove each scientist, group, or organization to take up a newly
fashionable field, we can learn something about the reward system
in their world. From the way that work in one sector stimulated
activity in another, we can learn about the linkages that bound the

sectors together. The opening section of chapter 4 recounts the responses to the laser of two professional societies, the Optical Society of America and the Institute of Radio Engineers (then merging into the Institute of Electrical and Electronic Engineers). For optical institutions, the laser was particularly opportune. It forged new connections between optical science and topics at the frontier of physics and thereby strengthened the legitimacy of optics at a moment when its position within the academic curriculum was being challenged.

The second section of chapter 4 describes the firms that began to offer commercial lasers starting in 1961. This section also recounts two key discoveries of 1961–1962: the method called Q-switching for concentrating the output of a ruby laser into a shortened pulse of heightened power, and laser action in Raman transitions. Hughes Aircraft Company was a central actor in both discoveries, and by examining events at that company we can show how important to the story was the interplay, characteristic of the period, between research into applications and research into basic laser phenomena.

Working scientists were drawn to the laser because of its scientific interest, because of the prospect of important phenomena to be discovered, because resources for laser work were available, and because publications on lasers commanded attention. In the field of gas lasers, examined in the third section of chapter 4, progress took several directions. New methods of exciting the gases into the lasing state were invented; additional gases were made to lase; laser tubes were improved; and understanding of the spectroscopy of the lasing gases was increased. These advances cross-fertilized each other to a point where the rapid pace of discovery itself became a motivation for research.

Gas was not the only type of lasing medium that was studied. Indeed, one crucial property of lasing is precisely that it can occur in many types of physical systems. This had the social concomitant that the laser boom spread from one community of specialists to another, infecting in turn glass scientists, semiconductor scientists, chemists, and hydrodynamicists. The semiconductor physicists were the most important group to enter in the early period. The fourth section of chapter 4 traces the race that developed among U.S. laboratories in 1962 to build a semiconductor laser, culminating in the simultaneous and independent discovery of the gallium arsenide diode laser at General Electric, IBM, and Lincoln Laboratory. The concluding section of chapter 4 looks briefly at the way mass circulation magazines portrayed these events. Their articles stimulated investors and thus closed the circle, as one sector after another of the high-technology establishment helped attract others onto the laser bandwagon.

By 1964, the euphoria was over, and a period of more sober research succeeded. It was carried on, however, at a high level of intensity. Worldwide laser research yielded a thousand papers a year in the open literature alone. In the United States, over the ten years from 1963 to 1973, laser R&D budgets rose from tens of millions of dollars to hundreds of millions.[27] Faced with such daunting numbers of individuals and institutions involved, volume of research done, and diversity of projects, I have fallen back on the method of case studies. Chapter 5 tells the story of five of the most important lasers of the 1960s and shows how they provided an impetus for new commercial companies. It then surveys five areas of applications research. Three were major applications of the 1960s: communications, interferometric measurement of length, and materials processing. The other two, laser fusion and laser isotope separation, hold their interest as applications that arose from the interaction of new technical ideas and changing national priorities.

Applications research was, of course, more directly influenced by social priorities than was device work. In addition, the two types of investigation were done by somewhat disjunct groups of scientists. I argue that the bridge between the two was the consensus among device scientists as to what properties make a laser desirable, a consensus that was heavily informed by the pronouncements of systems scientists. It remains a task for the future to explore whether the consensus arrived at within the specific conditions of U.S. research in the 1960s was any different than that formed in the Soviet Union, the nations of Europe, or Japan.

2

Masers[1]

The Ammonia Beam Maser

In 1950, the frontier for the generation of radio waves lay in the millimeter range. This was a culmination of an ascent up the electromagnetic spectrum to higher frequencies (or, equivalently, shorter wavelengths, since frequency and wavelength are inversely proportional) begun by the radio communications community during World War I. Previous markers had been wavelengths of tens of meters, achieved in the mid-1920s, and then meters, achieved by the early 1930s. By the late 1930s, radar researchers had worked with wavelengths of tens of centimeters and less, and in World War II, with the aid of powerful new magnetrons, they had made effective radars at single-centimeter wavelengths.[2]

Following the war, molecular microwave spectroscopists on the one hand and the military on the other spearheaded research into still shorter wavelengths. This juxtaposition is not as strange as it might at first seem. The discipline of microwave spectroscopy had blossomed as a result of wartime radar work. American scientists, who had begun by using the 10-centimeter (S-band) radar that the British had developed, had gone on to develop a 3-centimeter (X-band) device that, because of its shorter wavelength, gave better discrimination while using more compact equipment. To get still more accuracy, they next developed 1.25-centimeter (K-band) radar. Here, however, they met disappointment. The K-band beams penetrated only about 15 miles, as opposed to the 75–100 miles achieved by 3-centimeter radar. Studies were launched into the spectrum of water molecules to find out if the radar beams were being absorbed by atmospheric water vapor. This research confirmed water vapor to be the culprit and at the same time gave a powerful boost to microwave spectroscopic studies of molecules. After the war, the availability of 1.25-centimeter equipment encouraged further study. These radars were being discarded by the military, but they were "avidly sought out by physicists and chemists for microwave spectroscopy and [became] the envy of scientists in countries less blessed with discarded radars."[3] "So far as I know," one pioneer reported in 1948, "only one paper on microwave spectroscopy exists in prewar literature. . . . Yet in the brief period since the ending of World War II no less than 100 research papers . . . have appeared."[4]

Unfortunately, most molecules interact only weakly with radiation at the wavelengths, around 1 centimeter, that were available from radio sources. Their more intense spectroscopic lines are around 1 millimeter. The thermal sources then being used in infrared spectroscopy were not much good, though, for wavelengths longer than 0.3 millimeter. This provided a strong inducement for molecular spectroscopists to push microwave sources into the millimeter and submillimeter region. There was an additional incentive. The coherence and near monochromaticity of microwaves could make it possible to discern spectroscopic details with a resolution four orders of magnitude greater than that permitted by incoherent thermal sources. Spectroscopists had a reason, therefore, to want radio sources even in those wavelength regions already covered by thermal sources.[5]

Two founding fathers of microwave molecular spectroscopy who had started doing millimeter-wave spectroscopy by the end of the 1940s were Walter Gordy of Duke University and Charles H. Townes of Columbia University. They used electronic tubes with apparatuses that either produced, or isolated out, harmonics (integral multiples) of the tubes' fundamental frequencies. Gordy and his colleagues doubled and tripled the fundamental frequency of klystron tubes by passing their output through silicon crystals; Townes and his group used tapered waveguides to separate out the harmonics of magnetrons.[6] Both methods involved considerable sacrifice of power but still allowed power levels that were acceptable for spectroscopic experiments.[7]

The military was interested in millimeter systems because their more compact components would be useful in reducing the weight of guided missiles and in designing lightweight short-range radars for installation in tanks or submarine periscopes. Millimeter waves also promised greater secrecy for short-range communications. Precisely because at certain wavelengths they were so markedly attenuated by the atmosphere, they could not be heard by an enemy only slightly beyond their range. Finally, a new wavelength could bring an element of surprise, so that for a time at least the enemy would not know how to jam it.[8]

The laboratories of the Joint Services Electronics Program (JSEP) offered one group of facilities within which the research needs of the military and of spectroscopists could be pursued in parallel and with useful cross-fertilization. The Office of Naval Research, the Army Signal Corps, and the Army Air Force had established the JSEP in 1946 as a mechanism for providing joint funding to two wartime electronics research laboratories that they wished to preserve into peacetime: the MIT Radiation Laboratory, reorganized into the Research Laboratory of Electronics, and the Columbia Radiation Laboratory, within the Columbia University Physics Department. Before the end of 1946, a JSEP program was started at Harvard University also, uniting residues of the wartime Harvard Radiation Research Laboratory, which had specialized in electronic countermeasures, and the Harvard Officers' Training Course. In 1947, the Stanford Electronics Laboratory was brought in as the fourth JSEP laboratory.

The JSEP laboratories possessed charters that allowed their academic leaders broad latitude in research directions, subject to the condition that these directions were generally consonant with Department of Defense (DoD) interests. For the DoD, the JSEP laboratories were valuable because they promoted the science and technology from which future military breakthroughs might reasonably be expected to come, kept leading scientists in touch with the military's needs, and helped train future generations of scientific workers. For the universities, the JSEP laboratories provided equipment and money for faculty research and for the training of graduate students.[9] By 1950, Columbia Radiation Laboratory had one of the premier programs in DoD-related millimeter-wave research. At the same time, it housed Townes's work on molecular spectroscopy and the fundamental microwave studies of Willis E. Lamb jr. on the spectrum of hydrogen. At this time, it seemed right and normal to most U.S. scientists to pursue fundamental research and military applications together in view of the Soviet Union's achievement of an atomic bomb in 1949 and the recent outbreak of the Korean War.[10]

In 1950, the Electronics Branch of the Office of Naval Research asked Charles Townes to organize an Advisory Committee on

Millimeter Wave Generation. The committee's tasks were to identify promising lines of investigation, to stimulate research in the
field, and to advise the ONR on contract proposals.[11] Townes was
then in his middle thirties. A pioneer in microwave molecular
spectroscopy, he was also experienced in radar techniques. He had
designed and tested radar bombing systems, including K-band
systems, for the Air Force during the war as part of his work as a
physicist at the Bell Telephone Laboratories.[12] Townes recruited to
the advisory committee a small group of physicists and electrical
engineers from MIT, Stanford, Johns Hopkins, Bell Telephone
Laboratories, and the Naval Research Laboratory.

In April 1951, the committee had scheduled a daylong meeting
in Washington, D.C. Townes arrived the evening before the meeting. He awoke early the next morning and, so as not to disturb his
roommate, left his hotel room and settled on a bench in a nearby
park to mull over the basic problems the committee was facing.[13]
The conventional microwave sources—electron tubes, klystrons,
magnetrons, and traveling-wave tubes—all had striking disadvantages in the millimeter and submillimeter ranges. It was difficult to
fabricate them in dimensions tiny enough to produce these smaller
wavelengths and, even worse, difficult to get such diminutive
structures to dissipate the heat generated during operation.[14] Nor
had any of the unconventional methods proposed so far proven
feasible. A radically new idea was needed.

Molecular systems radiate at short wavelengths, and Townes's
thoughts turned to them. A molecular system exists in one of a
sequence of quantum-mechanical states of discrete energy, and it
can change its state from a lower to a higher energy level by
absorbing incoming electromagnetic radiation (see figure 2.1).
Conversely, when in a higher energy state, it can emit radiation by
one of two processes. The first is spontaneous emission. The
second, in which emission is induced, or stimulated, by incoming
radiation, is the exact reverse of absorption: In absorption, the
incoming radiation is appropriated by the system to raise its energy
level; in stimulated emission, the system moves to a lower level and
simultaneously adds energy to the incoming beam. If, as is generally the case, most of the molecular systems making up the aggregate body are in a lower state, there will be, on balance, net
absorption. But if an abnormal situation can be created in which
there is a population "inversion," that is, more systems in a higher
state, the incoming beam will itself be amplified.[15] Townes of course

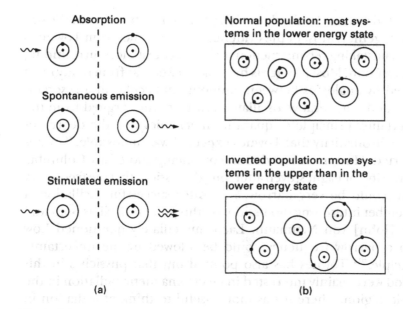

Figure 2.1
(a) Molecular absorption and emission processes. (b) Normal and inverted populations.

knew this, but he also knew that the expected amplification was rather weak.

Then, by his own account, Townes suddenly grasped a crucial idea. The upper-state systems could be enclosed in a macroscopic resonator designed to sustain radiation of just the frequency that they emitted. Then the stimulated emissions would be reflected back into the systems, inducing further emissions, with the result that radiation would start to increase. At high enough levels of radiation, the apparatus would break into self-oscillation, and at this point it would constitute the sought-for source of millimeter waves. Townes decided that deuterated ammonia (ND_3) would be a suitable molecule because it had a transition between energy levels that corresponded to a wavelength of 0.5 millimeter.[16] He expected that the apparatus would exhibit the sharply monochromatic output and coherence typical of all feedback oscillators.

Stimulated emission had been known to physicists for over 30 years, and "regenerative" oscillators, that is, oscillators with feedback, were well known to engineers. Why, then, was Townes's insight so novel? The answer appears to be that in 1951, physicists and engineers in the United States were not yet sufficiently acquainted with each other's territory to find it natural to put the two ideas together. Most of the evidence for this comes from Townes's own painstaking reconstructions of the circumstances surrounding his invention. Most physicists he spoke to were far from ready to be swayed by analogies between ammonia molecules in a resonant cavity and regenerative oscillators. Instead, they argued from the uncertainty principle of quantum mechanics that the degree of monochromaticity that Townes expected was impossible. Townes reported, "I never succeeded in convincing two of my Columbia University colleagues, even after long discussion, that the frequency width could be very narrow. . . . After successful oscillation, I remember interesting discussions on this point with Niels Bohr and with [John] von Neumann. Each immediately questioned how such a narrow frequency could be allowed by the uncertainty principle."[17] Townes has also pointed out that physicists in this period were mainly interested in electromagnetic radiation in the visible region, where it was more useful to think of radiation in terms of photons (discrete packages of energy) than in terms of waves. But the coherence of two streams of radiation is most naturally thought of in terms of waves. "Emphasis on the photon aspect of light deflected some physicists from coherent amplification," Townes writes. "In thinking about light itself, rather than microwaves, it may be that many electrical engineers would have not been any more concerned with coherence effects than were these physicists."[18]

The papers physicists wrote on stimulated emission before 1951 bear out Townes's analysis. They made no connection between stimulated emission and the technologies of coherent amplification or generation. For example, in a 1947 paper discussing an electric discharge through hydrogen gas, Willis E. Lamb jr. and Robert C. Retherford showed that a population inversion would probably obtain between some energy levels, so that there would be net "negative absorption" of microwaves at frequencies corresponding to transitions between those levels. But they did not break through the boundaries of the scientific problems they were considering to examine the potential of such negative absorption for amplification in electronics.[19] In 1950, in a widely noted experiment at Harvard University, Edward M. Purcell and Robert V. Pound demonstrated a population inversion, albeit a transient one, in a crystal of lithium fluoride placed in a magnetic field. The conclusion they emphasized, though, related to the foundations of physics, not to amplification: Their result made it possible to extend the concept of temperature to include negative values.[20]

Townes's achievement lay in part precisely in his ability to leap over disciplinary divisions and bring together ideas not customarily united. He was well positioned to do this. He was himself one of the founders of microwave spectroscopy and thus part of the group of physicists who worked in ranges of the electromagnetic spectrum where the wave picture was more useful than the photon one.[21] He was also deeply motivated to pursue the technology of coherent generation because of his field's needs for spectroscopic sources, because of his connection with the Columbia Radiation Laboratory, and because of his participation on the Advisory Committee on Millimeter Wave Generation. It is also worth noting that Townes directed the research of the Carbide and Carbon Chemicals Corporation postgraduate fellows at the Columbia Radiation Laboratory. This fellowship, first offered in 1948, grew out of the hopes of scientists at this unit of Union Carbide and Carbon Corporation that intense submillimeter waves could be used to catalyze processes in industrial chemistry. Townes applied the fellowship money to support projects on the generation of shorter wavelengths and spectroscopic studies of molecular structure. Meanwhile the fellowship, which had to be approved anew by the corporation each year, kept him and the Columbia Radiation Laboratory administration in touch with the interests and programs of Carbide and Carbon.[22] This connection was one means by which Townes kept abreast of some of the commercial possibilities of coherent short-wavelength radiation.

It is also to Townes's credit that he had the judgment to evaluate his idea as worthy of a sustained research effort. In the fall of 1951, he recruited Herbert J. Zeiger, a postdoctoral fellow who had just completed a thesis on molecular beams, to work on it, and shortly thereafter he arranged with graduate student James P. Gordon to make it the subject of a doctoral thesis. It would have been risky to stake Gordon's degree work entirely upon such a novel piece of apparatus, but Townes and Gordon figured that even if the device failed as a submillimeter wave generator, it would allow Gordon to glean new data on the ammonia spectrum.[23]

Shortly before Townes's moment of inspiration on the Washington park bench, Joseph Weber, a young professor of electrical engineering at the University of Maryland, had completed some calculations of his own with the same aim of yoking stimulated emission to the needs of electronics. Whereas Townes was a physicist who had learned microwave engineering through wartime radar work and had deepened his knowledge through subsequent scientific and technical projects, Weber came to his union of physical and engineering concepts by an opposite route. He had started as a microwave engineer and served from 1945 to 1948 as head of the electronic countermeasures section at the Department of the Navy's Bureau of Ships. Here he had ample chance to learn the technological importance of highly sensitive amplifiers at microwave and millimeter wavelengths. Countermeasures, which are directed against an enemy's radar, require receivers that employ such amplifiers to detect the radar waves.[24]

Weber left the navy in 1948 and enrolled as a PhD student in physics at Catholic University. Quantum mechanics was part of his first-year studies, and he later recalled that as soon as he heard Karl Herzfeld's lectures on stimulated emission, he realized that such emission might be applicable to microwave amplification. Weber now began to ponder how a practical stimulated-emission amplifier might be realized.[25] The Purcell-Pound result supplied the ingredient he needed. It offered a way to create a population inversion that would allow for net stimulated emission and hence amplification. The Electron Tube Research Conference, an informal yearly meeting at the cutting edge of electronics, was to be be held in Ottawa, Canada, in June 1952. Weber submitted a paper in early 1952.[26]

In two respects, Weber's work was orthogonal to that of Townes. First, whereas Townes's initial interest was in a generator of radiation, Weber's was in an amplifier. The idea of feedback, so essential to Townes's conception, was therefore not vital to Weber, and he did not include it. Second, whereas Townes planned for a physical separation of the molecules in states of higher energy from those of lower energy, Weber proposed to keep the two kinds together and to impose an inversion of populations by adding energy. The numbers Weber got out of his calculations indicated such minimal performance that he decided not to try to reduce his idea to practice.[27] But his discussion of these ideas in his conference paper (published in 1953) and in other talks brought to the consciousness of his audience the technical interest of population inversion.[28]

In contrast, Townes did decide to build a device and therefore considered the character of an apparatus through which the physical principles might find practical expression. Three components were necessary: a source of ammonia molecules, a focuser to separate molecules in the upper energy state from those in a lower state, and a resonator. The resonator would of course be some sort of a microwave cavity, since the radiation of interest was in the microwave to far-infrared region; its precise dimensions would be determined by the wavelength of the transition being amplified.

The sine qua non for oscillation was that adequate power be released to the cavity by the ammonia molecules flowing through it. This required, first of all, that as large an intensity as possible of upper-state molecules be focused into the cavity. Townes decided to adapt for his ammonia molecules a novel type of high-intensity focuser that two German physicists had just devised for atoms.[29] Second, it required that the cavity have good energy retention (not be too "lossy"). Detailed calculations made in the fall of 1951 showed that it would be hard to make a cavity for 0.5-millimeter radiation that would retain energy well enough to permit oscillation.[30] Townes decided to shift from the 0.5-millimeter wave in deuterated ammonia to the 1.25-centimeter wavelength in ordinary ammonia. Here a better cavity from the viewpoint of energy retention could be achieved; moreover, there were good off-the-shelf detectors and waveguides at 1.25 centimeters but not at 0.5 millimeter. By the same token, however, there were already good off-the-shelf generators of 1.25-centimeter radiation. Townes's decision thus shifted the project from one of pushing back the frontiers into the submillimeter region to one of demonstrating a novel principle of generation in an already-achieved region (figure 2.2).[31]

Zeiger and Gordon began their work in early 1952. In February 1953, Zeiger completed his postdoctoral fellowship at Columbia and left for a position at MIT's Lincoln Laboratory. By this time they had built a cavity that was lossless enough to allow the beam to amplify signals fed into it, although it was not adequate for oscillation. Gordon continued on alone under Townes's direction. By June 1953, he had most of the equipment in place. Much of the next six months was consumed in debugging, but Gordon was able to demonstrate amplification by year's end.

There was still insufficient power to allow oscillation. Gordon now removed the annular slit through which the beam had been entering the cavity, leaving a hole with diameter two-thirds that of

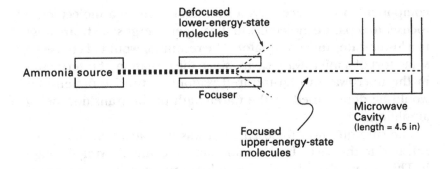

Figure 2.2
Schematic diagram of an ammonia beam maser.

the cavity wall. This increased by tenfold the number of molecules admitted into the cavity. In early April 1954, with this change and with other smaller improvements, he obtained oscillation with a continuous power output of about 0.01 microwatt, "the first time that energy has been obtained continuously from a molecular resonance."[32] The Columbia group called their apparatus a "maser" because it provided microwave amplification by stimulated emission of radiation. Gordon was now in a position to fulfill his thesis requirements with both spectroscopic measurements on ammonia and the maser itself, a device that was at once an amplifier and an oscillator (figure 2.3).

The ammonia beam maser was the first apparatus to run by stimulated emission, but it was not the first in which transitions between energy states of microscopic systems were pressed into use for technological ends. Already in the late 1940s, scientists had stabilized microwave oscillators on absorption transitions. The National Bureau of Standard's ammonia absorption clock, completed in 1949, pushed this line of work a stage further. The cesium beam clock, another type of quantum resonance clock being developed at the National Bureau of Standards, at MIT, and at Great Britain's National Physical Laboratory, was based upon a

Figure 2.3
Charles H. Townes (left) and James P. Gordon flank the second ammonia
beam maser in a photograph taken at Columbia University in 1955. The side
has been removed to show the maser's internal structure. In the rear, T. C.
Wang stands beside the first maser. (Courtesy Smithsonian Institution)

transition between two hyperfine energy levels in the cesium
atom.[33] Some of the applications of ferrites, a new class of highly
magnetizable, low-electric-loss materials that was then being vig-
orously exploited, also depended upon resonance transitions. And
early results that would convert the phenomenon of nuclear
magnetic resonance into a practical method for measuring the
earth's magnetic field to high accuracy were just being reported.[34]
The maser was thus an ingenious and important advance within a
field that already existed, one that within a few years would be given
the umbrella name of "quantum electronics" by its practitioners.

As an oscillator, the output of the first maser was weak, but it
conformed to Townes's expectations and confounded the skeptics
by its exceptional monochromaticity. Exactly how pure was its
frequency? To answer this question, Townes and his group
constructed a second maser to compare with the first. By early
1955, they were able to report a spectral purity of 4 parts in 10^{12} over

intervals of one second, and 1 part in 10^{10} over intervals of one hour. These figures made the maser an attractive candidate for a precise frequency standard or an atomic clock. At this time, the capabilities of the cesium clock were not yet clear.[35] Meanwhile the National Bureau of Standards had brought the ammonia absorption clock to an accuracy of 1 or 2 parts in $10^{8,}$ and Tokyo University had squeezed out another factor of 10. It did not appear, however, that the ammonia clock could be refined much further.

High-precision clocks and microwave frequency standards had a number of potential applications. Precise atomic clocks could be used to determine whether the astronomical "constants" remained fixed or changed with time, and they could also form the basis for tests of the theory of general relativity. Standards were important for microwave spectroscopy. Beyond their scientific significance, they had important commercial and military uses. Accurate frequency standards were needed for both ground stations and aircraft in the system of long-range radio-guided air navigation that was then in its final stage of development under U.S. Army Signal Corps sponsorship. Highly stable shock-resistant frequency standards were also part of the electronics package that was mounted in guided missiles.[36]

Research on ammonia gas masers soon began to spread to a handful of other laboratories in universities, government installations, and industry. The vagaries of personal contact and the interest and support of the military, particularly the Army Signal Corps Engineering Laboratory, both played a role in determining the pattern of diffusion. At the Jet Propulsion Laboratory of the California Institute of Technology, the management at the Electronics Research Section sent a newly hired scientist, Walter H. Higa, to Columbia University in the fall of 1954 to learn the maser art. JPL wanted to know if the ammonia maser could be developed into a transportable frequency standard for use on guided missiles.[37]

At Stanford University, physicist Willis E. Lamb jr. started to work on the theory of the ammonia beam maser. Lamb had been a colleague of Townes at the Columbia University Physics Department and the Columbia Radiation Laboratory before he moved to Stanford in 1951. He had heard about the maser then, but he had not expected it to work. Its success now goaded him into studying its mechanisms.[38] Meanwhile, an electrical engineering student at the Stanford Microwave Laboratory, John C. Helmer, read Gordon,

Zeiger, and Townes's 1954 paper and arranged to do an experimental thesis on the maser for his doctorate. He went to Columbia University for a month in January 1955 to get instruction in the hardware and then to Lamb to gain a deeper understanding of the theory. Helmer received his degree in 1956 and joined the electronics firm of Varian Associates. He brought his maser expertise as part of his qualifications, and he spent his first two years, as a member of the tube division research group, working on a Signal Corps contract on the ammonia maser's properties.[39]

Among his affiliations, Charles Townes was a consultant to the Microwave Standards Section of the National Bureau of Standard's Central Radio Propagation Laboratory. In April 1954, Townes wrote to Harold Lyons, chief of the Microwave Standards Section and the leading figure behind the NBS atomic clock program, to call his attention to the ammonia maser: "I have become more and more impressed with the possible use of this molecular oscillator as a frequency standard."[40] Lyons was deeply interested. His work on atomic clocks and on frequency standards based on the resonant transitions of molecular systems was part of an overarching commitment to a program of replacing the physical standards then in use with atomic standards. Lyons promptly inaugurated a program of research on the ammonia beam maser.[41]

In mid-1955, Lyons moved from the Bureau of Standards to the Hughes Aircraft Company's Research and Development Laboratories. Hughes Aircraft Company had been established as a division of the Hughes Tool Company in the mid-1930s to support industrialist Howard R. Hughes's hobby of aviation. During World War II, the company had become a defense contractor, and in 1945 it had begun to pursue the business of military electronics. By 1950, Hughes Aircraft was selling radar fire-control systems for jet fighter-interceptors and had major contracts for the development of air-to-air guided missiles. It was the Korean War, however, that initiated the company's dramatic takeoff. Within months of the conflict's start, orders for fire-control systems mushroomed into the thousands. Hughes's postwar technical leaders, Simon Ramo and Dean Wooldridge, had implemented a policy of hiring top-ranking electronics scientists and using them to tackle R&D projects that were at once tough enough to deter other aircraft companies and central enough to military needs to guarantee a long-term market. This policy fit neatly into the Department of Defense's emphasis on funding the development of state-of-the-art weaponry. Hence

Hughes Aircraft's fortunes rose. Employment went from about 1,000 in 1948 and 6,000 in 1950 to about 16,000 in 1952 and over 20,000 in 1955.[42]

At the time that Lyons joined Hughes, the company's research activities were being reorganized. Previously, the Research and Development Laboratories had embraced production design, systems engineering, and field engineering as well as research and development in guided missiles and radar. In 1954, however, the Hughes Research Laboratories, comprising the Electron Tube Laboratory and the Microwave Laboratory (into which Lyons was hired), were separated off organizationally. Hughes Research Laboratories had more than 600 employees by late 1956 and was responsible for developing the advanced components Hughes needed for its systems. The leaders of the laboratories, director Andrew V. Haeff and Microwave Laboratory head Lester C. Van Atta, appreciated the importance for the company of the interaction between components scientists and systems scientists, and a strong coupling between the two groups was to remain a hallmark of Hughes research. But they also thought it essential to deepen the scientific base for components research. They therefore established several new programs, one of which was the program in atomic and molecular resonances, located within the Atomic Physics Section of the Microwave Laboratory, that Lyons came to head.[43]

Hughes was interested in the ammonia beam maser because, as a frequency control system, it had potential applications in navigation and missile guidance. It appeared to them, moreover, that variants of the maser might yield oscillators in the important millimeter and submillimeter regimes or might be developed into low-noise microwave amplifiers. Lyons therefore brought a maser program to Hughes when he migrated there, and he soon succeeded in using his old contacts with the Signal Corps Engineering Laboratory to negotiate a wide-ranging and open-ended contract for gas maser development work.[44]

Townes was also a consultant to Airborne Instruments Laboratory, a small company that had its origins in a group set up under the Office of Scientific Research and Development during World War II to design submarine detectors for aircraft. At war's end, while other research groups were metamorphosing into university or government laboratories, the scientists at Airborne Instruments had recruited some of their wartime collaborators at the Harvard

Radio Research Laboratory and the Columbia Radiation Laboratory and formed a commercial firm.[45] This laboratory now undertook to manufacture the Columbia University ammonia beam maser, and in 1955 they secured a contract with the Air Force Rome Air Development Center for studies leading to a stable, flyable maser frequency standard.

Bell Telephone Laboratories began its maser program by hiring James P. Gordon, Townes's student, at the beginning of 1955. Like many other U.S. electronics laboratories, industrial and government, Bell Laboratories had grown rapidly following the onset of the Korean War. Employment had risen from roughly 5,000 employees in 1950 to more than 9,000 in 1955. The proportion of personnel engaged in military work grew from 12% in 1950 to 23% in 1955,[46] but other inputs were also contributing to the expansion. Telephone consumption was zooming. The United States had gone from having fewer than 20 phones per 100 people in 1945 to more than 30 in 1955.[47] Taking into account other forms of communication, such as television and military links, Bell Laboratories leaders foresaw "almost unlimited growth" in the demand for communications service.[48]

To meet the projected demand, Bell Telephone Laboratories had mounted a major effort in millimeter wave communications starting in 1950.[49] In early 1955, it would begin to work on satellite communications, under the promptings of John R. Pierce, the director of Electronics Research. It was into Pierce's area that Gordon came, recruited by Rudolf Kompfner, a Viennese-born architect-turned-engineer who was then head of a department within Electronics Research.[50] Kompfner was interested not so much in the maser's potential as a frequency standard as in its promise as a low-noise microwave amplifier. In this role, it might be useful as a component in a system for receiving signals relayed from satellites. While the rest of Kompfner's group worked on traveling-wave tubes, an amplifier Kompfner had invented while working for the British Admiralty during the Second World War, Gordon's project was to refine and build ammonia beam masers.[51]

The RCA Laboratories in Princeton offer a different scenario that also merits discussion. RCA was an important early player in the use of molecular transitions for frequency control. Robert H. Dicke, professor of physics at nearby Princeton University and a consultant at the laboratories, was an intellectual leader in this effort. Like Townes, Dicke had been immersed in radar during the war; he had joined the MIT Radiation Laboratory in 1941, immediately after obtaining his PhD. Like many others worldwide, Dicke had left radar research determined to apply to science the microwave techniques that had been developed for military purposes. Whereas Townes's interest led him to become one of the founders of microwave spectroscopy, Dicke used radar techniques in other ways, such as the measurement of fundamental properties of the hydrogen atom. Well before the Columbia University maser papers were published in 1954 and 1955, Dicke's research had led him to a number of topics intimately related to maser physics. One was the use of light to change the populations of energy levels in an ensemble of molecular systems. This was the technique of "optical pumping" that would eventually be applied to the production of a population inversion between two energy levels in a maser material.[52] Another such topic was stimulated emission.

A rough map of the path Dicke took between the study of hydrogen and coherent emission begins with line narrowing. In order to "narrow" spectroscopic lines, an investigator must reduce the effect of any extraneous influence that might alter the frequencies of the emitting systems and thus cause the total radiation to spread over a broader frequency range.[53] Dicke needed narrow spectral lines in order to read the subtle features he and his students were exploring, features such as the interaction of the electron in the hydrogen atom with its nucleus. Line narrowing could be fostered by lowering the density of particles in the gas under study, so as to eliminate the line broadening due to interparticle collisions. Low particle densities, however, meant a low signal-to-noise ratio in the results. To improve the signal-to-noise ratio, Dicke adapted a procedure introduced in 1952 into the study of nuclear spectroscopy by H. Y. Carr and E. M. Purcell of Harvard.[54] In the nuclear version, one or more radiofrequency pulses are used to raise a group of nuclei into excited spin states. After the pulse, the ensemble spontaneously sends out a pulse of its own, one that carries information on the nuclear energy levels and is easier to detect because the exciting pulse is no longer around to contribute background noise. The radiated pulse has some of the coherence properties of stimulated emission because of the common influence on the nuclei of the preparation pulse. Dicke translated this method from radio frequencies to microwave frequencies and from nuclear spin states to the states of atoms.[55]

Having achieved this result in the realm of instrumentation, Dicke was impelled by scientific curiosity to study coherent radiation more generally by the methods of quantum mechanics.[56] The radiating atoms, he showed, are locked together not only by the pulses that initially prepare them in coherent excited states but by the radiation that bathes them—radiation they both emit and experience.[57] As a result, pulses of microwaves (and, by implication, optical pulses) could be created that were much more intense and more narrowly collimated than ordinary light. Dicke called these "superradiant" pulses. Thus, quite independently of Townes and by a very different path, Dicke was also moving toward the technology of generation of coherent short-wave radiation.[58]

A three-sided relationship involving Dicke, the Signal Corps, and RCA Laboratories mediated between Dicke's ideas and their translation into technology. Dicke's university group had been partly supported by Signal Corps contracts since at least 1950. As a result, he was well informed about the work of the Corps and its contractors in such areas as frequency standards and clocks. Dicke's center of interest as a physicist was in fundamental problems, but he also found enjoyment and relaxation in inventing. In his notebooks, suggestions for practical gadgets such as home heating devices, automotive accessories, and telephone systems are interleaved with designs for scientific instruments and ideas for new experimental techniques. His consultancy for RCA made him feel obliged to translate some of his ideas into patents or into projects for the laboratory staff. But the ideas RCA wanted, "in the science of radio-frequency energy level transitions in atoms and molecules and its application to technology," were themselves conditioned by the fact that RCA held a number of Signal Corps contracts in these areas.[59]

Narrow line widths were important for frequency standards, for example, and RCA conducted a three-year Signal Corps project to narrow line widths based on Dicke's ideas on coherent emission.[60] One of Dicke's students, James P. Wittke, joined the RCA David Sarnoff Research Center in September 1955 and pursued, on company funds, Dicke's maser-based proposal called the "hot grid cell." The idea was to apply temperature and voltage gradients in such a way that the creation of a small group of high-energy-state molecules, the separation of these molecules from those of lower energy, and the stimulated-emission process could all take place within one self-contained cavity. Such a sealed-off cell, liberated from the throughput of ammonia molecules that was a feature of the Columbia University maser, was interesting for many purposes, but Wittke's focus was on its utility as a microwave amplifier. Like Gordon at Bell Telephone Laboratories, Wittke ran his one-person project within a group whose other members were working on low-noise traveling-wave tubes for amplifiers.[61]

Looking back over the period to the end of 1955, we observe that maser research in the United States originated with university scientists: Charles Townes of Columbia University and Joseph Weber of the University of Maryland.[62] From this core it spread to industry, other universities, and government laboratories. Personal contact loomed large in the early period, but the crucial point is that it was contact with all sectors of the U.S. research establishment. Academic physicists were tied not only into academic networks but also into industry, as consultants, and into government, as contractors, members of committees, and consultants. Students graduated not only into university positions but into industrial and government research laboratories. The initiation of maser work at the National Bureau of Standards was facilitated by Townes's status as a consultant to NBS. Dicke's ideas were exploited at RCA because of his consultancy with the Sarnoff Center. Townes's student, James Gordon, went to Bell Telephone Laboratories. Dicke's student, James Wittke, found a place at RCA.

Military funding played a vital role. The Columbia University work was done under the auspices of the Joint Services–supported Columbia Radiation Laboratory. The research at the National Bureau of Standards was internally funded, but the level of that funding was largely set by representatives of the military.[63] When he moved to Hughes, Lyons worked under a Signal Corps contract. Airborne Instrument Laboratory's program was sustained by the Air Force's Rome Air Development Center. RCA was aided by Signal Corps money. At Stanford University, both Lamb and Helmer were on a Signal Corps contract.[64]

The military contribution went beyond financial support. By way of example, the Signal Corps Engineering Laboratory in 1956 took a variety of actions to stimulate work on the ammonia beam maser. It issued requests for proposals and then actively worked to identify and solicit the U.S. research groups it wanted to recruit. It organized conferences for the exchange of information and the coordination of effort among its contract maser teams. It monitored the quality of the work done, expedited the purchase of laboratory equipment, and served as a clearinghouse for information.[65] It reviewed the state of the field and called to the technical community's attention the directions of advance it saw as desirable, directions such as better sources of excited molecules, better focusers, improved vacuum systems, and sealed-off units.

The total number of research programs on gas masers in the United States through 1955 was small, and the resources being expended were minimal. There were about ten institutions involved, and the sizes of the programs ranged from one-person efforts, as at Bell Telephone Laboratories, to a group of four or five at Columbia. Nevertheless the gas maser work, along with work on other technological uses of molecular resonances, such as atomic clocks and nuclear resonance, created a pool of expertise that would be essential to maser science and, ultimately, the development of the laser.

Solid-State Maser

As a microwave amplifier, the ammonia beam maser had one outstanding advantage: It added almost no noise to the incoming signal. It therefore made possible the construction of microwave receivers orders of magnitude more sensitive than conventional receivers. The ammonia beam maser had, however, two major shortcomings. First, the amplification *bandwidth*, that is, the band of frequencies on either side of the ammonia line's center for which there would be appreciable amplification, was very small.[66] Second, the degree to which the frequency of the center of the line could be changed, or *tuned*, by the imposition of magnetic or electric fields was also small, so that circuit designers would pretty much be stuck near the 24,000-megacycle frequency of the ammonia transition line.[67] These twin drawbacks of narrow bandwidth and poor tunability limited the effort invested in research into amplifiers based on ammonia beam masers. And where amplification was the primary goal, as for example at RCA, other types of gas devices were preferred: The hot-grid ammonia cell Wittke was studying at RCA promised a bandwidth 10 times greater than that of the beam maser.[68]

The way to improve bandwidth and tunability would be to abandon the vibrational transition in ammonia gas in favor of other types of quantum transitions in other species of molecules, ions, or atoms. One attractive class of transitions consisted of electron spin transitions (either ferromagnetic or paramagnetic) and nuclear spin transitions. These could enhance tunability because the effect that changing the magnitude of an external magnetic field has on the frequency of a paramagnetic ion in a solid, for example, could be as much as a thousand times greater than the comparable effect of an electric or magnetic field on an ammonia molecule. Using solids instead of gas would also mean enormous increases in power, because, to use the same example, the concentration of paramagnetic ions in a solid can easily be 10^5 times greater than the number of ammonia molecules in a beam; this increased power could be translated into increased bandwidth.[69]

Townes had some of these considerations in mind during his 1955–1956 sabbatical, when he undertook a collaborative research at the University of Paris on solid-state masers. Townes had in fact planned to use the year to decide whether to remain in microwave physics or strike off in a new research direction.[70] But when he arrived at the École Normale Supérieure in the fall of 1955 for the first leg of his leave, a new piece of physics claimed his attention. A former student, Arnold Honig, who was working with the French physicist Jean Combrisson on electron paramagnetic resonance, told him that arsenic ions embedded in silicon crystals at liquid helium temperatures exhibited the unusually long relaxation time of about 16 seconds. Hearing this, Townes recognized that this made arsenic-doped silicon a candidate for a solid-state maser material: An arsenic ion in an upper spin state would remain there long enough to allow useful energy to be extracted by stimulated emission. During the 1955 Christmas season, Townes sketched out a number of possible systems. Some of these emulated the ammonia beam maser in that a continuous ribbon of doped silicon traversed a microwave cavity, and microwave radiation entering the cavity was continuously amplified by it. Some simpler schemes used a silicon crystal fixed in the cavity to permit intermittent amplification. Both types differed from the ammonia maser in one respect: In the original maser, the ammonia molecules in the upper energy state had been physically separated from the molecules in the lower state, and mainly the upper-state molecules entered the cavity. In a solid, this separation was impossible, and instead a population inversion was established.[71]

Townes persuaded Combrisson and Honig to begin experimental work in early 1956. What they envisioned was a pulsed maser using paramagnetic ions such as arsenic in silicon, phosphorus in silicon, or manganese in hydrated silicon-fluoride. To obtain a population inversion, they applied a pulse of microwave energy.[72] By early spring of 1956, when Townes was due to move on to Tokyo for the second lap of his sabbatical, the three investigators had gotten power outputs up to one-third of those they needed—not enough for oscillation, but enough to convince Townes that this was a promising direction. Meanwhile Townes had spawned a related effort at Bell Telephone Laboratories. He had visited the laboratory during the winter to obtain experimental crystals and to tell James Gordon and others of his enthusiasm for this line of research. Shortly after Townes wound up his own attempts to realize his ideas and left for Japan, Gordon teamed up with George Feher from the Semiconductor Research Group to see if they could create a workable two-level solid-state maser.[73]

At about the same time, but independently of Townes, the MIT physicist Malcolm Woodrow P. Strandberg began to proselytize for the solid-state maser. Strandberg, like many protagonists of quantum electronics, had worked on microwave radar during World War II. His stint was at the MIT Radiation Laboratory and in Great Britain with the Royal Air Force. After the war, he entered the field of microwave spectroscopy, and then, in the early 1950s, he became interested in paramagnetic resonances. Strandberg took as his purview the study of the energy states in, and the interactions among, the atomic, molecular, and ionic systems making up gases, liquids, and solids, including the design and construction of equipment to explore the properties of these systems.

Strandberg was affiliated with both the Research Laboratory of Electronics (RLE) and the Department of Physics at MIT. The RLE was interdepartmental. It was initially designed to facilitate interaction between physicists and engineers; later, other specialties were added. RLE was also, as we have seen, funded by the Joint Services Electronics Program through a contract administered by the Army Signal Corps. This circumstance put RLE faculty in communication with Department of Defense monitors and through them with the DoD's research interests and needs.[74] Because of his RLE affiliation, his Radiation Laboratory background, and even more because of his consultantships with defense firms and his participation in DoD committees, Strandberg maintained an active interest in the technological relatives of his laboratory apparatus. He kept up the tradition, emanating from the wartime Radiation Laboratory, of physicists who took on the engineering of military or civilian technologies "unspoiled" by engineering training or preconceptions.

Strandberg recently recalled that he had been moved to champion solid-state masers in part by the program on ammonia beam maser amplifiers being undertaken by Lyons's group at Hughes Aircraft Company. This program, in his view, was wrong-headed, since ammonia masers lacked sufficient bandwidth for any useful amplifier application. Solid-state masers, on the contrary, had potential "attributes . . . that should make the electrical designer's eyes sparkle."[75] In 1956, as Strandberg worked to get into gear his own research program on quantum-mechanical amplifiers, undertaken jointly with electrical engineering professor Robert Kyhl, he also spoke and wrote on the subject, attempting to enlist scientists and engineers in the solid-state cause.[76]

Strandberg's proselytizing was, on one occasion at least, brilliantly successful. On May 17, 1956, he concluded an MIT colloquium on paramagnetic resonance with a few remarks on the virtues of solid-state masers. In the audience was Nicolaas Bloembergen, a Dutch-born associate professor at Harvard's Division of Engineering and Applied Physics and an authority on solid-state physics. Bloembergen was puzzled. The gas maser was a very pure frequency standard; a solid-state maser would not have this purity. Why did Strandberg advocate it? He questioned Strandberg after the seminar, and Strandberg told him that what he had in mind was an entirely different application, that of ultra-low-noise microwave amplification. The fruitfulness of Strandberg's idea struck Bloembergen. He was a consultant at this time to the solid-state research group at Lincoln Laboratory, a research center that had been set up in 1951 to provide the Department of Defense with state-of-the-art radar, communications, and computers. Clearly ultra-low-noise amplifiers would be useful in radar and communications.

Bloembergen discussed Strandberg's ideas with Benjamin Lax, leader of the solid-state group. But both Strandberg and the paper of Combrisson, Honig, and Townes, which the Lincoln group now showed Bloembergen, had discussed two-level masers. These had two drawbacks in that they were by-and-large pulsed devices and were restricted to solids with abnormally long relaxation times. A device that evaded the restriction to special materials , and that could run continuously, would clearly be more useful.[77]

For several weeks, Bloembergen mulled over possible schemes. Then on June 12 and 13, the "good idea" came. He would use a molecular system with three unequally spaced energy levels and would pump systems from the lowest to the highest level, supplying energy at a rate that would make the number of systems in the highest state (level 3) equal to the number in the lowest state (level 1) (see figure 2.4). Then either the middle state (level 2) would have a smaller population than the other states, in which case he could get stimulated emission from level 3 to level 2, or it would have a larger population than the others, in which case he could get stimulated emission from level 2 to level 1.[78] The fact that the pumping transition was divorced from the masing transition in the three-level maser allowed continuous action and lifted the limitation to materials with long relaxation times.

Behind this insight, so quickly achieved, lay Bloembergen's intimate knowledge of two postwar fields, nuclear magnetic resonance and electron spin resonance, that were rich in concepts and methods applicable to masers.[79] Bloembergen had moved from the Netherlands to Harvard University early in 1946, just after Edward M. Purcell, Henry C. Torrey, and Robert V. Pound had first demonstrated the existence of nuclear magnetic resonance. In this phenomenon, a substance—the Harvard group used paraffin—absorbs energy from a radiofrequency signal, with a resulting change in the number of nuclei that populate various nuclear spin energy levels. Bloembergen had become Purcell's doctoral student and had done research that culminated in an important paper by Bloembergen with Purcell and Pound on "Relaxation Effects in Nuclear Magnetic Resonance Absorption."[80] Subsequently, Bloembergen had extended his research to include electron paramagnetic resonance phenomena, in which microwave frequencies are used to alter the distribution of populations in electron spin energy levels.

Transient population inversions were well known in nuclear magnetic resonance phenomena, and stimulated emission was a familiar idea. Purcell and Pound had introduced the concept of negative temperature in 1951 in the context of research on nuclear magnetic resonance. Their work, as we have already noted, influenced Joseph Weber as he formulated his thoughts on the maser. Another recently demonstrated phenomenon with which

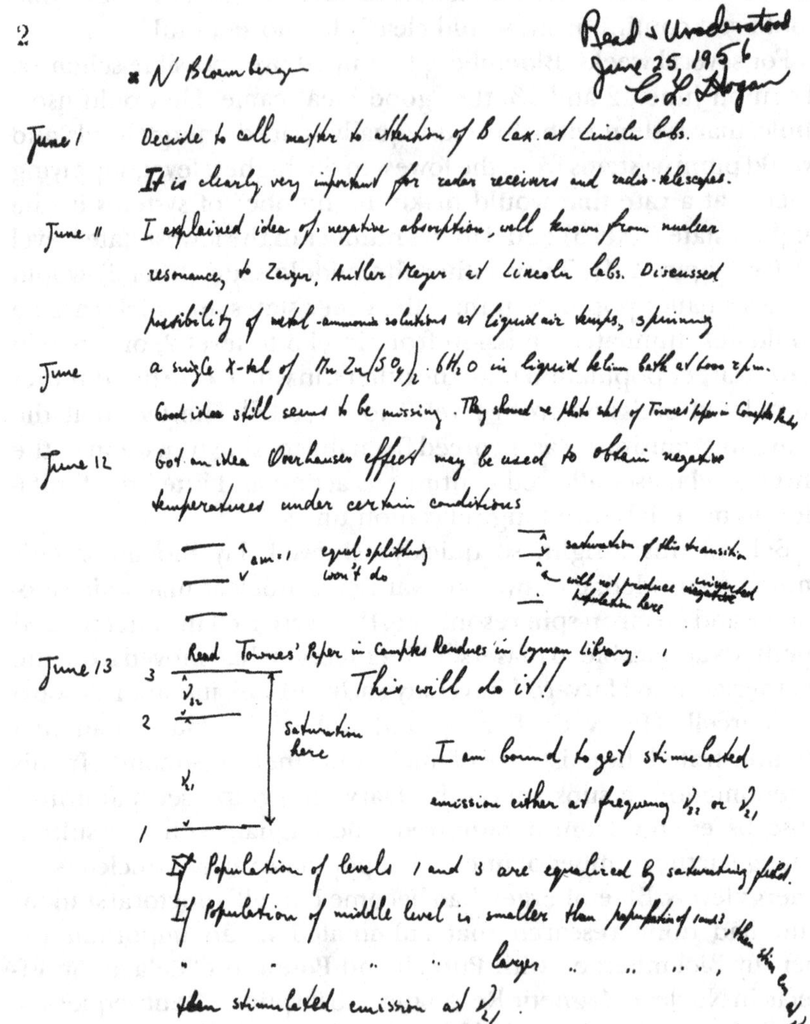

2

N Bloembergen

Read & Understood
June 25, 1956
C. L. Dogan

June 1   Decide to call matter to attention of B Lax at Lincoln Labs.
It is clearly very important for radar receivers and radio-telescopes.

June 11   I explained idea of negative absorption, well known from nuclear
resonance, to Zeiger, Butler Meyer at Lincoln Labs. Discussed
possibility of metal-ammonia solutions at liquid air temps, spinning

June   a single x-tal of Mn Zn(SO₄)₂ · 6H₂O in liquid helium bath at low r.p.m.
Good idea still seems to be missing. They showed me photo stat of Townes' paper in Comptes Rend.

June 12   Got an idea Overhauser effect may be used to obtain negative
temperatures under certain conditions

—— am=1 equal splitting   —→ saturation of this transition
—— v   won't do   —← will not produce inverted
——   population here

June 13   3 ——┐   Read Townes' paper in Comptes Rendues in Lyman library.
            v₁₂ ┊   This will do it!
         2 —— ┊ saturation
              ┊ here   I am bound to get stimulated
         x₁   ┊   emission either at frequency v₁₂ or v₂₁
         1 ——┘

If Population of levels 1 and 3 are equalized by saturating field.
If Population of middle level is smaller than population of level 1, then
   "   "   "   "   " larger "   "   "   "
then stimulated emission at v₂₁.

Figure 2.4
Bloembergen started a notebook on 22 June 1956 to record the steps he had
followed to that date to arrive at his proposal for a three-level solid-state
maser, and he continued to record new ideas as he developed them. This is
page 2. (Courtesy N. Bloembergen)

Bloembergen was familiar was the Overhauser effect, in which populations in two electron spin levels are equalized, thereby creating an enhanced population difference between two nuclear spin levels.[81] Here was an outstanding example of engineering a population difference between two given energy levels by enforcing equal populations on two other levels. Although Bloembergen's idea was different in significant respects, the Overhauser effect was nevertheless part of the mental equipment that enabled Bloembergen to work out his three-level scheme.

But were there any natural materials that could be pumped in the way Bloembergen was suggesting? At first blush, it appeared to him that transitions from level 1 to level 3 were all forbidden.[82] He overcame this first, and for him highest, hurdle on June 14 when he realized that these normally forbidden transitions would become allowed if one set the direction of the externally imposed magnetic field at an angle to the internal crystalline field axis. In a week of intensive work, Bloembergen selected possible materials—gadolinium ethyl sulfate, dilute nickel fluosilicate, and perhaps nickel-magnesium fluosilicate—and wrote down a set of governing equations. In the following week, he worked out some of the technological features of his idea, such as the magnitude of the power output, methods for modulating the output, and the optimal crystal size and device temperature. On June 25, he handed the first draft of a paper to a university typist. He now started to work on translating his ideas into a radiotelescope receiver that could detect the microwave radiation of interstellar hydrogen.[83]

Meanwhile, the Bell Telephone Laboratories team of Gordon and Feher had found evidence of masing in a doped silicon sample. Feher presented preliminary results at the Electron Tube Research Conference in Colorado in July. 1956. Shortly thereafter, Rudolf Kompfner, now Director of Electronics Research at Bell, proposed as one possible though "speculative" future goal for the company's maser program "[to] make a continuously-acting amplifier."[84] To this end, Kompfner approached H. E. Derrick Scovil, a member of the technical staff. Scovil was in the Solid State Device Development area rather than in Kompfner's group, but the two men knew each other from Oxford University, where Kompfner had worked before he had come to Bell Laboratories and where Scovil had taken his doctorate in electron paramagnetic resonance physics. Kompfner now suggested that Scovil look at the problem of making a continuously acting solid-state maser. On August 7, Scovil submitted a memorandum proposing a solution; he had also prepared a draft of a paper for the *Physical Review*.[85]

The gist of Scovil's scheme was identical with Bloembergen's, although the construction of the argument was quite different. Instead of starting with a general discussion, Scovil elected to develop an argument for gadolinium ethyl sulfate as a representative crystal. This was a salt that he knew thoroughly from his thesis work, and his familiarity with it speaks out in the comfort with which he proceeded, first demonstrating the way in which the problem of forbidden transitions could be overcome, then tackling the question of relaxation interactions, and moving on to a rough estimate of the expected amplification.[86] As against Scovil's insight into his representative salt, however, it may fairly be said that Bloembergen's notebook and paper showed more generality and a richer treatment of the physics issues, and that Bloembergen had also proceeded substantially farther with the technological details.

By this time, rumors of Bloembergen's work had begun to reach Bell Laboratories, while Bloembergen was learning that something was afoot at Bell. Bloembergen wanted to patent his maser, and he began to worry that he may have been too forthcoming to colleagues with preprints and oral information. Bell was worried about a potentially nasty situation, with recriminations over the provenance of ideas and, further in the future, questions of patent infringement. On September 7, Bloembergen accepted an invitation to New Jersey to present his results to the Bell Laboratories staff. Scovil later recalled that he had not heard about Bloembergen's three-level maser when he had composed his own memorandum. But "when Bloembergen came here it was quite clear that he had had the same idea, that in fact he had certainly had it before I did. So I did not send my paper for publication." Instead, Bell Laboratories negotiated an agreement to take out the patent for Bloembergen, while the two parties agreed to a friendly rivalry as to who would be the first to reduce the three-level maser to practice.[87] At Harvard, Bloembergen commenced work with research fellow Joseph O. Artman and graduate student Sidney Shapiro, using first a nickel-zinc fluosilicate crystal from Lincoln Laboratory and later potassium cobalticyanide doped with chromium ions.[88] At Bell Laboratories, Scovil teamed up with George Feher and electrical engineer Harold Seidel to pursue a gadolinium ethyl sulfate maser.[89] At the same time, at Lincoln Laboratory,

Benjamin Lax asked two staff members, James W. Meyer, a physicist with a background in both microwave paramagnetic resonance and radar, and Alan L. McWhorter, an electrical engineer with semiconductor expertise, to undertake a three-level maser project.[90] Townes had by this time returned to Columbia University and had started a solid-state maser project of his own.[91]

Willow Run Laboratories, a facility owned by the University of Michigan, also began solid-state maser work in 1956. In the mid-1950s, Willow Run's main contract, for the Army Signal Supply Agency, was Project Michigan, a comprehensive study of methods of battlefield surveillance encompassing radar, acoustics, seismic detection, infrared sensing, and naked-eye observation. Willow Run had a staff of about 600, part University of Michigan faculty and students and part scientific workers without faculty status. As a university affiliate, Willow Run could offer its employees the opportunity to take courses and obtain advanced degrees.[92]

In 1955, Project Michigan engineer Weston E. Vivian had started a small effort in the unconventional area of passive microwave surveillance. This was, in a sense, a compromise between radar, which actively sends out microwaves and records their reflections, and infrared sensing, which passively registers natural emissions at infrared wavelengths. Passive microwave sensing seemed to have a potential for expanding surveillance activities; it was also close enough to the forefront of technology to make it a suitable subject for Vivian's thesis research. From data on earlier work, Vivian calculated that he would need microwave receivers of extraordinary sensitivity. He had been following the development of masers and, sometime in the spring or summer of 1956, obtained funding for a maser program at Willow Run.[93] Laboratory administrators assigned the program to a new staff member, physicist Chihiro Kikuchi, who had been hired into the Infrared Laboratory in June 1955 to help initiate solid-state research. Maser research complemented Kikuchi's program of studying crystals by measuring their microwave absorption properties.[94]

At the time, Willow Run lacked equipment and expertise for research on electron paramagnetic resonance. This constraint interacted with scientific considerations in shaping Kikuchi's thinking as his maser studies took definite form in late 1956. He was, first of all, spurred by a remark by Project Engineer Newbern Smith that there was no point in starting yet another maser program unless it embodied a novel approach. He therefore took a fresh look at crystals that might serve as hosts to masing ions. He decided to stay away from the popular hydrates, such as the nickel-zinc fluosilicates, that Bloembergen was then pursuing,[95] because the strong magnetism of the hydrogen nuclei exerted a disturbing influence on the active ions. He also rejected Bell Laboratories' choice, gadolinium ethyl sulfate, both because of its nuclear magnetism and because its crystal properties made it unsuitable for amplifying radiation at wavelengths shorter than about 10 centimeters.[96] The substance that most attracted Kikuchi initially was chromium trichelate, a compound he had learned about from his former colleagues at the Naval Research Laboratory.[97] His technicians, however, had trouble growing chromium trichelate. Pink ruby, which is chromium-doped aluminum oxide, was commercially available, however, and had many of the same advantages for amplification below 10 centimeters as the trichelate. In January 1957, Kikuchi switched to ruby, with chromium trichelate reduced to the role of backup. He obtained a sample of pink ruby from the University of Michigan's mineralogy department.[98]

One important parameter in the manufacture of a maser is the angle that the externally imposed magnetic field makes with the crystalline axis. The separations among the energy levels, and therefore the frequencies the maser can amplify, depend on this angle. At this time, the favored angle was about 15°, but to solve for the energy levels at 15° required computing facilities. Kikuchi decided to survey the shape of the energy levels using 54°46′; for this special angle, the equations simplified radically, and an analytical solution was possible.[99] He soon realized that the angle had another virtue: It would permit amplification at a frequency (9,300 megahertz, corresponding to a wavelength of about 3.2 centimeters) at which the laboratory machinists had experience through their radar work.[100] Further calculations convinced Kikuchi that 54°46′ might in fact allow a workable maser. The ruby program now became the main string in the Willow Run bow.[101] The work went slowly, though, because of the need to set up liquid helium facilities from scratch.

In December 1956, while Kikuchi and his team pursued the ruby maser, Scovil, Feher, and Seidel at Bell Laboratories became the first team to operate a three-level maser. They used gadolinium ions in a salt of lanthanum ethyl sulfate. Their work was aided by a bit of "molecular engineering." The extent to which gadolinium ions amplify electromagnetic energy depends on the amount of population inversion that can be achieved between level 2 and level 1, the two levels involved in the masing transition. This in turn depends on, among other things, how rapidly ions pumped from level 1 to level 3 drop down to level 2. The Bell team took advantage of the fact that their salt included the element cerium as an impurity. As solid-state physicists, Scovil and Feher knew that the cerium would interact with the gadolinium ions to increase the rate at which gadolinium ions at level 3 relaxed downward to level 2.[102] Scovil and Feher's knowledge of the behavior of paramagnetic ions enabled them to design molecular properties into the maser just as, for example, an electrical engineer like Harold Seidel designed microwave cavities and waveguides.

The Bell team operated their maser as an oscillator but not as an amplifier. In the spring of 1957, McWhorter and Meyer at Lincoln Laboratory succeeded in operating a maser, made out of potassium cobalticyanide doped with chromium, as both an amplifier and an oscillator, at a wavelength of about 10 centimeters. Shortly thereafter, in collaboration with a research engineer in a local company, they measured an upper limit to maser noise, proving experimentally that the cavity maser had by far the lowest noise of any amplifier then known.[103] Townes and his group put a maser of potassium cobalticyanide material, with a wavelength of 3 centimeters, into operation in August 1957. Bloembergen, Artman, and Shapiro operated their 21-centimeter maser, also made of potassium chromium cobalticyanide, on December 7.[104] It was only on December 20, 1957, that Kikuchi and his coworkers—George Makhov, John Lambe, and Robert W. Terhune—demonstrated maser action.[105] Their ruby maser, however, made the other maser crystals immediately obsolete. Ruby was an off-the-shelf material and thus readily available. It was both tough and easier to adjust for tuning purposes.[106]

The three-level solid-state maser was an ultra-low-noise, tunable, continuously acting microwave amplifier of respectable gain. This was bound to attract the attention of electrical engineers. In October 1956, even before the Scovil-Feher-Seidel demonstration, John R. Pierce at Bell Laboratories wrote to British scientist Richard Q. Twiss, "Right now there is great hope in this country that masers may be the real clue to low noise in the microwave range."[107] At about the same time, Stanford electrical engineering professor

Hubert Heffner wrote to Nicolaas Bloembergen, "A small group of us here at Stanford have been interested for some time in starting a research effort in microwave solid-state devices. . .. It would appear that [the three-level maser] is the first really attractive scheme."[108] Sometime in the latter part of 1956, senior electrical engineering faculty member Dean Watkins suggested to Anthony E. Siegman, who was just then finishing his doctoral thesis and preparing to start an appointment as acting assistant professor, that he take up solid-state masers as a research field.[109] Siegman began in early 1957 to construct a three-level maser using potassium chromicyanide and to initiate wide-ranging studies in maser theory, maser materials, and novel device design.[110] In the fall of 1957, Heffner and a colleague, Glen Wade, themselves started a small effort on two-level solid-state masers.

At this time, a reformation of the electrical engineering curriculum was already in progress in the United States, with the aim of diversifying the field and strengthening its scientific content.[111] The absorption of the maser into the armamentarium of the engineer gave impetus to this movement. It required instructors to introduce their students to quantum-mechanical energy levels and level lifetimes, the elementary laws of spontaneous emission, and the distinction between spontaneous and stimulated emission.[112]

But it was not only undergraduates who found that they needed to learn quantum mechanics. Granted that many of the engineers who took up masers had prior experience with physics. Siegman, for example, had taken his bachelor's degree at Harvard's Division of Engineering Science and Applied Physics and had chosen to do his master's degree, at UCLA, in applied physics rather than engineering.[113] Robert W. Hellwarth, soon to collaborate on a major contribution to the theory of masers, was the son of an electrical engineer and was planning to be an engineer himself, but he fell in love with physics in high school and took a double major in electrical engineering and physics when he went to Princeton.[114] Other engineers, however, both graduate students and faculty, now found that they needed to catch up on modern physics quickly.

Amnon Yariv had just finished a master's thesis on traveling-wave tubes in the Electrical Engineering Department at the University of California, Berkeley, and was shopping around for a PhD topic when he attended the July 1956 tube conference and heard George Feher describe the work at Bell Laboratories on the two-level solid-state maser. Yariv was attracted. In part it was a matter of personality: Feher was a physicist with style and imagination. But in part it was the conceptual richness that Yariv sensed in the field.[115] He had had no previous training in quantum mechanics, but he solved this by spending the 1956–1957 academic year entirely on graduate physics courses in quantum mechanics, group theory, solid-state physics, and microwave resonance spectroscopy. In Scovil's interdisciplinary team of engineers and physicists at Bell Laboratories, the physicists taught the engineers quantum mechanics and the engineers taught the physicists about electronic devices.[116] Richard H. Pantell at Stanford University recalled that he participated in a study group with two other engineers in the mid-1950s to learn quantum mechanics.[117]

As engineers entered maser research, they brought engineering science and engineering perspectives to the field. They contributed specifically engineering insights to the field of maser noise, they studied optimal methods for the integration of maser components into systems, and they supplemented the original solid-state maser, which used a cavity resonator, with a traveling-wave maser.

Consider the problem of noise. Noise can be understood as a manifestation of the fluctuation of a quantity from its average value. In an amplifier, the power put out at any given instant will usually differ from the average power put out over a long period of time. As a result, the amplifier seems "noisy," and the level of noise is determined by a function specifying the probability that the instantaneous power output has a value different from the average power output.[118] We could also interpret amplifier noise by making a model of the physical amplifier in terms of a fictitious "equivalent" electronic circuit, and this was a standard procedure in engineering. The circuit elements are seen as adding "noise power" to the signal power entering the circuit, and the noisiness of the amplifier is defined by a "noise figure" that is a ratio of two ratios: the ratio of signal power to noise power entering the amplifier, and the ratio of signal power to noise power exiting the amplifier.

The maser posed a cornucopia of intriguing questions for noise theorists. For example, could the concept of noise figure be applied to a maser, since the equivalent circuit for the maser had to contain "negative" resistances? How could the results of communication theory, which were formulated for purely classical systems, be generalized so as to be applicable to a quantum system such as the maser? Was there a lower limit below which the noise of a maser could not be reduced? All previous amplifiers had been treated by classical theory, which gave zero for the lowest theoretical noise. This had no practical significance because in any previous amplifier the noise was always large. But the maser could not be treated by a purely classical theory and, in addition, seemed to have a noise so low that the question of the existence of a lower limit became practically interesting. And if, in fact, quantum theory were to predict an irreducible minimum for the maser's noise, then wouldn't the other "classical" amplifiers have noise minima also, since, after all, the physical universe is at bottom a quantum universe?

Faced with such an array of interesting questions, both physicists and engineers were quick to take up theoretical and experimental research on maser noise. MIT engineers Hermann A. Haus and Richard B. Adler contributed redefinitions of the concept of noise figure.[119] Columbia University physicists Charles Townes and Robert Serber wrote on fluctuation theory and the uncertainty principle.[120] It would be false to the facts to visualize the engineers entering under the banners of noise figures and equivalent circuits and the physicists marching in under the colors of fluctuation theory and the quantum-mechanical uncertainty principle. Nevertheless it is clear from little asides in the scientific literature that engineers and physicists knew they had something to teach each other about noise, and also something to learn from each other. Thus, as late as the February 1963 International Quantum Electronics Conference, Haus and J. A. Muller proselytized the attendees: "As optical masers become more widely used . . . the utility of simple theoretical formalisms increases. In engineering applications of amplifiers, equivalent circuits have found widespread use because they give a visual representation of equations and the physical phenomena for which they stand."[121] And physicist Eugene I. Gordon wrote a tutorial article about engineering treatments of noise in microwave transmission lines, so that those "who do not have a background in microwave theory and techniques . . . may find a very modest introduction to an extremely well developed store of computational techniques which are applicable to optical masers. This may save them the trouble of inventing their own."[122]

Another quintessentially engineering contribution was the traveling-wave maser. First, it reflected a characteristic engineering goal: to improve the operating parameters of the maser amplifier. Second, the branch of knowledge that underlay it—traveling-wave tube structures and their properties—was a part of engineering science, specifically of microwave electronics engineering. Third, it was principally engineers who brought the traveling-wave maser into existence.

The first masers were all "cavity" devices. The masing material emitted radiation during the time it was resident in a microwave cavity. This was true for Townes's ammonia beam maser, in which the ammonia molecules streamed into the cavity to do their radiating and then passed out of it again. It was true for the two-level maser Combrisson, Honig, and Townes attempted and the one Feher and Gordon brought to successful completion. And it was true for Scovil, Feher, and Seidel's three-level maser. The cavity functioned as a resonant reflector. Both the cavity and the maser crystal were tuned to the weak signal that was to be amplified, the maser crystal through imposition of a varying external field and the cavity through adjustment of its geometry. The signal wave thereby exerted the strongest possible effect upon the masing material as it passed through it again and again because of reflection from the cavity walls.

There were some disadvantages to the use of a cavity, however. One was that every time the static magnetic field was changed to tune the maser to a new signal frequency, the cavity also had to be retuned. Another was that the amplification bandwidth could only be improved at the expense of the gain, because the product of the bandwidth and the square root of the gain is a constant for a cavity. Still another disadvantage was instability. A good amplifier operates just at the limit of self-oscillation, that is, just before it switches from operation as an amplifier of weak incoming signals to operation as a generator of its own signal. In a cavity, it is easy to jump above the limit and (from an amplifier viewpoint) go unstable.[123] All three of these difficulties can be avoided by using a traveling-wave maser. In a traveling-wave tube, there is no reflection. The signal wave passes down the material only once. The structure of walls, wires, and dielectric and metal elements is arranged, however, so that the wave travels extremely slowly (about 1% of its speed in air), so that it has time to squeeze the maximum response out of the masing material.

The traveling-wave maser was an obvious idea for electronics engineers, whose lowest-noise amplifier to that time was a traveling-wave tube in which a slow wave interacted with a beam of electrons. Traveling-wave masers were being discussed even before the first three-level maser was operated—for example, by Henry Motz at Oxford University and Rudolf Kompfner at Bell Laboratories.[124] They became even more interesting after the first cavity masers were operated, as researchers instigated systematic attempts to get the best possible performance out of maser amplifiers. When Anthony E. Siegman and his graduate students at the Stanford Electronics Laboratory had wrung the optimal possible performance out of their cavity maser, they terminated that program and concentrated exclusively on trying to build a traveling-wave device.[125] McWhorter and Meyer at Lincoln Laboratory turned to traveling-wave masers after they completed their pioneering study of noise in the three-level cavity maser.[126] The maser group at Bell Laboratories also worked on traveling-wave masers and, in fact, created the first operational system toward the end of 1958. The structure they used for slowing the wave was based on original ideas by Harold Seidel, modified and improved by Robert W. DeGrasse, another newly hired Stanford-trained electrical engineer.[127]

Maser Applications

Solid-state masers were conceived and put into operation at a time when the need for low-noise amplifiers in the microwave region was growing. Radar, which operated in these wavelengths, was a protean technology and the core of a substantial business, and low-noise receivers could potentially increase the range of radar by a factor on the order of 5. The military used radar in planes, ships, and ground installations. As jet aircraft moved into civilian use, civilian aviation became another important market. In 1958, the Federal Aviation Agency required that radar be installed at the largest U.S. airports. Counter-radar measures also created a demand since the technology of countermeasures relies crucially on the ability to detect weak enemy signals.

Low-noise receivers were also of lively interest to radioastronomers, who explore objects in the solar system and beyond by studying the radio waves they emit. Astronomers had just begun moving down the wavelength scale, from emissions in the meter-and-longer range to emissions of tens of centimeters and then to centimeter lengths. In 1951, H. I. Ewen and Edward M. Purcell of Harvard became the first to detect the 21-centimeter line emitted by galactic hydrogen. By 1954, workers at the U.S. Naval Research Laboratory had broken through to radioemissions at 3.15

centimeters.[128] At these shorter wavelengths, however, researchers ran into a number of technical problems, one of which was that their receivers used triodes or traveling-wave tubes as amplifying elements. Because of the intrinsic noise of these amplifiers, heroic efforts were required to extract information about the very weak celestial signals.[129]

Space applications provided another growing source of demand. Low-noise ground receivers were useful for tracking missiles and satellites, receiving telemetered messages from space vehicles, and charting the natural electromagnetic emissions in space, thus delineating the environment within which space systems would operate. Receivers mounted in space probes or satellites were in surroundings of much lower noise than earth-bound receivers, and this made minimizing their own noise figures more desirable than ever.

Demand for low-noise amplifiers helped shape the initial enthusiasm that greeted the solid-state maser. In 1956 and 1957, a wave of optimism washed over systems engineers and funders alike. "Much is expected . . . of the solid-state amplifier," wrote one commentator, pointing out that the internal noise of a maser receiver should be a factor of 40 less than that of traveling-wave tube microwave receivers.[130] Airborne Instruments Laboratory began a one-page article, which it ran as an advertisement, "Since the solid-state maser was first announced, the imagination of engineers has been fired by the vision of finally beating the thermal noise limitation of microwave receivers with which they have lived so long." The Electron Tube Research Conference featured masers prominently in its June 1957 sessions.[131]

The application of masers to microwave receivers did not, however, take place in a fixed technical environment. Two ongoing developments affected it: the invention of other kinds of low-noise devices, and the continuing research on masers themselves, which simultaneously improved their properties and made ever widening circles of engineers acquainted with their remaining limitations. One of the more formidable limitations, for example, was the fact that masers had to be run at liquid helium temperatures, that is, a few degrees above absolute zero. This was a nuisance in scientific contexts such as radioastronomy. It was still more serious for military applications such as field radar that needed to be simple enough for soldiers to service and rugged enough to function in the hostile environments of war or space. The newer low-noise amplifiers did not have as low a noise figure as the maser, but neither did they have as severe a low-temperature requirement. As time went by, masers began to lose out to their competitors for all but the most sensitive experiments. By 1963, H. E. Derrick Scovil could remark that the most important "application" of masers might well have been the stimulus they had given to the invention of more convenient low-noise amplifiers.[132]

In radioastronomy, it was often the maser scientists, rather than the astronomers, who took the initiative in developing masers appropriate to astronomical requirements. In this regard, they functioned as instrument makers in the service of the astronomers' research programs. The instruments they made, however, were still novel enough to win recognition as achievements within physics.

Bloembergen, as we saw, started laying plans for a 21-centimeter maser amplifier in June 1956, even before he had mailed his "Proposal for a New Type Solid State Maser" to the editors of the *Physical Review*. By mid-July, he had already enlisted research fellow Joseph O. Artman and graduate student Sidney Shapiro in the project. Bloembergen was well aware of the excitement over the 21-centimeter line from interstellar hydrogen. He had received his doctorate from Edward Purcell and had often discussed his maser proposal with Purcell in the course of its development. Bloembergen gave the 21-centimeter maser highest priority.[133] At this point, of course, he was not only trying to create an amplifier that would be useful to astronomers, but he was also competing with Bell and Lincoln Laboratory to come in with the first operating three-level solid-state maser.

Charles Townes was even closer to radioastronomy. In 1955, as he had started his sabbatical, he had seriously considered taking it up as a research field. The promise he found in solid-state masers during his Paris stay had deflected him, but when he returned to Columbia in the fall of 1956, he immediately set in motion a research project to make a maser receiver for a radiotelescope. For a test site for research at decimeter and centimeter wavelengths, Townes needed a telescope with a large antenna; he therefore approached the Naval Research Laboratory, whose 50-foot antenna was at that time the largest in the nation. NRL's program of measurements of 3-centimeter radiation from the planets was then achieving exciting results, in particular an unexpectedly high figure for the temperature of Venus. Townes and the two graduate students he brought into the project, Joseph A. Giordmaine and Leonard E. Alsop, therefore chose to design a 3-centimeter maser. The Columbia maser would enable NRL to repeat its measurements with more sensitivity.

Townes had started his program on the maser receiver before he heard about Bloembergen's three-level scheme; once he had had a chance to evaluate Bloembergen's proposal and make some initial trials, he abandoned his own ideas for two-level continuous-wave solid-state masers in favor of the Harvard three-level type.[134] In August 1957, Townes's group operated a three-level potassium chromium cobalticyanide maser at 3 centimeters. In 1958, after Kikuchi's team had demonstrated the ruby maser, the Columbia group switched to ruby, and this was the crystal used for the NRL telescope.

Kikuchi's work on the ruby maser also had the effect of bringing the University of Michigan into maser astronomy. Michigan was just then building a radioastronomy program around Fred T. Haddock, who had come to the Electrical Engineering Department in February 1956 from the Naval Research Laboratory's radio-astronomy group. The core facility was to be a new 85-foot telescope; that equipment and a receiver for solar radioastronomy were the principal focus of Haddock's attention. He also had a substantial background in radar, however, and shortly after joining the Michigan faculty, he became head of a committee evaluating a Willow Run radar project.[135] Haddock, who was a gregarious man, quickly developed wide contacts at Willow Run through his consultantship and through other electrical engineering faculty members with connections at Project Michigan. Thus he was in a position to learn of the maser work from Vivian before 1956 was out, and he jumped at the chance it promised for a more sensitive microwave receiver.[136] For the maser group in turn, Haddock's interest meant the oppor-tunity, through the publication of astronomical papers, to get more of their results into the unclassified literature. Within Kikuchi's group, Robert W. Terhune headed the team that put together a maser package for the 85-foot telescope.[137] The Willow Run maser team and University of Michigan astronomers would make a number of finds with this maser radiometer, including the first measure-ment of radioemissions from Mercury and one of the first from Saturn.[138]

Well before the Michigan measurements got going, however, substantial progress was made on other low-noise devices. Traveling-wave electron tubes were rejuvenated by a breakthrough that cut their minimum noise level in half.[139] Still more important was the solid-state parametric amplifier, which was different in principle from both electron-tube amplifiers and maser amplifiers. In electron-tube amplifiers, a weak electromagnetic input signal strengthens itself by feeding on the energy of a stream of electrons with which it is in interaction. In a maser amplifier, the input signal is amplified by energy received from excited but neutral molecular systems. In a parametric amplifier, it is a circuit element (a variable capacitance or inductance), itself continually pumped with energy by an outside high-frequency oscillator, that gives up energy to the signal.

In the United States, two scientists, independently and at opposite ends of the country, took hold of the idea of a microwave parametric amplifier in 1956 and 1957. Both were stimulated by the maser amplifier. Walter H. Higa at the Jet Propulsion Laboratory in Pasadena, California, had been working on masers since late 1954. In the fall of 1956, he wrote an internal memorandum pointing out that the underlying reason why masers had low noise was they did not use charged particles, as conventional electron-tube amplifiers did. But parametric amplifiers, an idea with roots in the nineteenth century, also had no need for charged particles and should also have low noise. Higa gave a talk on his idea at the June 1957 Electron Tube Research Conference in Berkeley, California. He developed the equations for a microwave parametric amplifier but did not give a specific embodiment of such a device.[140]

The other scientist, Harry Suhl of Bell Telephone Laboratories in New Jersey, started from a particular embodiment. Suhl had been studying the processes that occur when high-power electromagnetic radiation interacts with a class of magnetic materials known as ferrites. The advent of the three-level maser inspired him to look into the way in which these ferrite processes might be used to amplify microwaves.[141] Suhl recognized that a ferrite amplifier would share with the maser the feature of low noise, since both worked by means of quantum transitions in atoms, rather then by the flow of electrons, and both could be operated, if necessary, in cold environments.[142] Suhl's idea was speedily recognized as a variant of a parametric amplifier. Suhl's amplifier was, in itself, never of importance, although a working device was constructed to his specification by a Bell colleague.[143] What was important was the stimulus it and Higa's proposal gave to the invention of other kinds of parametric amplifiers for the microwave spectrum. Starting in 1958, the trade press reverberated with announcements of new parametric amplifiers. RCA made public a germanium diode amplifier in July 1958, and Bell Laboratories announced a silicon dioxide diode amplifier. In August, Zenith published news of a new electron-beam parametric amplifier; here again Bell followed shortly with a different version.[144] The noise temperatures for these devices were in the range from 15 to 100°K.[145]

Although the maser had a much lower noise temperature, 10°K and below, it had drawbacks as a systems component that seemed increasingly significant. To begin with, for applications in radio astronomy, the maser receiver had to be mounted at the center of the antenna dish if full use were to be made of its low-noise property (figure 2.5). This was awkward because masers operated at a temperature a few degrees above absolute zero and were maintained at this temperature by being encased in a pair of Dewar flasks; the inner flask held the maser surrounded by a bath of liquid helium (4.2°K), while the outer flask was filled with liquid nitrogen (77°K) to insulate the helium bath. These flasks had to be refilled daily. Joseph Giordmaine, who was working as a graduate student on the NRL-Columbia experiment, recalled, "This [equipment] was a major undertaking in a lot of ways . . . even the logistics of attaching and operating the liquid helium Dewar at the focus of an antenna . . . which is being pointed everywhere. . . . Loading this, you had to climb up a ladder about 20 feet in the air and pull up . . . a Dewar full of liquid helium to the focus of the antenna and install that there. The Dewar contained enough liquid helium for a run of about eight or ten hours."[146]

It also turned out that the maser was hard to tune and was easily detuned. Its very sensitivity was a drawback in the absence of an antenna specifically suited to it, because the maser registered noise that the antenna picked up from the ground, and the amount it registered would vary if the antenna dish were rotated so as to track the celestial object under study.[147]

Of course, parametric amplifiers had their own drawbacks. Because the technology was so new, it was often difficult to get good diodes. Moreover, the introduction of superconducting magnets

Figure 2.5
Leonard E. Alsop works on the Columbia University maser at the focus of the
50-foot parabolic dish at the Naval Research Laboratory. (Courtesy
Smithsonian Institution)

after 1959 led to important improvements in maser systems. Nevertheless the initial excitement over masers soon abated as applications research became increasingly a matter of making hard-nosed economic and technical comparisons between the maser and other available low-noise amplifiers.[148] This is what happened in astronomy. A 1966 review of radio observatories found that 11 of them were using masers; 65 used paramagnetic amplifiers, even though the average noise temperatures of the systems were four times as great; and 12 used amplifiers with still higher noise figures. The importance of the noise figure varied with the kind of experiment being done. The choice of amplifier also depended on the region of the spectrum being examined and the nature of the radioemission (for example, whether it was continuous or discrete). But it must also be acknowledged that the choice of amplifier reflected factors such as the remoteness of the installation, the sophistication of its other equipment, and the engineering skill of its scientists—all of which worked against the maser because of its greater complexity.[149]

For radar, the maser presented still other problems. The heart of the maser, consisting of the paramagnetic crystal and its resonant cavity, was small, but the entire unit was large—the size of a desk as compared to a traveling-wave tube's shoebox dimensions—because of the ponderous magnet surrounding the flasks. This was a particular drawback for radar designed for airplanes or space vehicles. Another difficulty was that masers installed in radar systems, although they were included as part of the receiver to help detect the reflected radar pulse, were also affected by the outgoing pulse. Some of the outgoing power invariably leaked through to degrade the population inversion and limit the maser's amplifying ability.[150]

Hughes Aircraft Company is one firm that began by exploring a wide role for masers in radar and ended by giving them a restricted one. The lion's share of Hughes's business in 1958 was aircraft electronics, or "avionics," for fighter-interceptors. That business was being threatened by the Defense Department's increased emphasis on missiles. Hughes in 1958 had in its pocket a major contract to provide the F-108, a new fighter that North American Aviation was building, with electronics systems. Hughes was promising that the radar incorporated into its systems would have a greatly expanded range. One of its arguments for the superiority of the manned fighter to the unmanned ballistic missile became the very possibility of this novel radar, with a range several times greater than that of conventional radar.[151]

The major technical development behind Hughes's argument was the maser. Lyons's Atomic Physics Section had by 1957 extended its research from the ammonia beam maser to both two- and three-level solid-state masers. As part of a many-faceted program, it had duplicated Bell Laboratories' gadolinium ethyl sulfate maser in 1957 and Kikuchi's ruby maser in 1958.[152] Thus the technical expertise of Lyons's group supported the company's marketing strategy even while that marketing strategy elevated the importance of the company's maser research.

The maser radar program at Hughes started in early 1958 as a collaborative effort of the Atomic Physics Section and the company's radar engineers. Within the Atomic Physics Section, Theodore H. Maiman, a physicist-engineer who had joined the company in 1956, created a new maser design that eliminated the external magnet in favor of a magnet inside the Dewar, thereby reducing the weight from 5,000 to 25 pounds.[153] The radar group designed innovative solid-state switches that could shield the maser from most of the transmitter power leakage without spoiling the low noise temperature of the receiver unit, and the maser group increased the band of frequencies over which the maser could operate.[154] But in the end, despite these and other improvements, Hughes chose parametric amplifiers for its airborne system, reserving maser radar for specialized ground installations.

The new science of radar astronomy also demanded highly sensitive receivers. Radar astronomy differs from radio astronomy in that it works not with the natural radioemissions of celestial bodies but with waves bounced off them from earth. One highly publicized early radar experiment, using first masers and then parametric amplifiers, was made at Lincoln Laboratory.

Lincoln Laboratory had been founded in 1951 with the charge of inventing and developing systems that could warn of attacking bombers; this was part of the U.S. response to the 1949 revelation that the Soviet Union had become the world's second nuclear power. By the mid-1950s, Lincoln Laboratory had been given the added mission of developing systems for warning against ballistic missiles. These tasks called for a marriage of radar, communications, and digital computers, and Lincoln Laboratory teams worked on radar in all its manifestations, from portable battlefield units through mammoth missile tracking installations.[155] The maser's potential for radar was one of the reasons why Benjamin Lax had originally organized a maser program, and Lincoln therefore became a pioneer in maser radar.[156]

In 1958, Lincoln scientists decided to take advantage of their exceptional radar, maser, and computer facilities to mount a program in radar astronomy. Experiments starting right after the war had obtained radar reflections from the moon. Lincoln Laboratory decided to try for Venus, 100 times farther away. The laboratory built this program around three of its technical achievements: first, the high power of its 84-foot antenna, put into operation in 1957 as an experimental prototype radar and also used to track Sputnik; second, the ability of its computers to separate out the faint pattern of radar echoes from the radio signals of extraneous origin that would swamp them; and finally, a maser with 1-meter wavelength that laboratory scientist Robert H. Kingston had designed and was just putting into operation. The low noise of Kingston's maser would boost the sensitivity of the receiver by a factor of 4.

The Lincoln team recorded Venusian radar data in February 1958, and in March 1959 it announced that it had successfully extricated radar echoes from its data.[157] (By contrast, the first of the radiotelescope experiments, the collaboration between NRL and Columbia, started collecting data in April 1958.) On the Lincoln scientists' second shot in August 1959, they substituted for the maser a paramagnetic amplifier, which needed cooling only to liquid nitrogen temperature, rather than liquid helium temperature. It was less complex but, with the aid of advances in other components, gave a substantially equivalent noise figure.[158]

The Lincoln Laboratory case exemplifies the competition between masers and other low-noise amplifiers in radar. It also reveals the symbiosis that existed between military and astronomical applications. The Venus radar shot made use of an antenna built for the military purpose of detecting missiles. As a converse example, at the Naval Research Laboratory, members of the radio astronomy group were called upon to apply their know-how to the development of a radio sextant that would make it possible for ships to steer by solar radioemissions in bad weather.[159]

Bell Laboratories presents a situation in which the symbiosis expanded to include an application that was as much civilian as military: satellite communications. And the scientific work with which satellite technology was intertwined at Bell Laboratories led to a result of Nobel prize-winning stature: the discovery by Arno A. Penzias and Robert W. Wilson of astronomical evidence for the "big bang" theory of the origin of the universe.

In 1954, a group of Princeton engineers invited Bell scientist John R. Pierce to give a talk on some aspect of space research. Pierce was stimulated to make some calculations on communications satellites.[160] At this time, AT&T, in collaboration with the British Post Office and Canadian groups, was just laying the first transatlantic telephone cable.[161] Pierce argued that a satellite link could well be the next step in transoceanic communications. He examined in his article both passive satellites, which would merely reflect incoming radio waves back to earth, and active satellites, which would receive incoming signals and transmit outgoing ones. Both types might be placed in either geosynchronous orbit, about 22,000 miles above the surface of the earth, or in a low-lying orbit at about one-tenth that altitude. One factor that would determine the practicality of these schemes was the sensitivity of the receivers at the ground terminal. A more sensitive receiver would reduce the transmitter power required. The most sensitive receivers then available used traveling-wave tubes as amplifiers, and Pierce based his calculations on these. He estimated that reflecting spheres in geosynchronous orbits—perhaps the simplest and most flexible technology—would require too much power. Low-lying spheres would need about 100 kilowatts.[162]

Other industrial laboratories also started to look at communications satellites in the mid-1950s,[163] but it was only after the launch of Sputnik, in October 1957 and the Explorer I satellite at the end of January 1958 that the idea really took hold. On the one hand, the sine qua non for the technology—functioning satellites—was now at hand. On the other hand, a number of ancillary technological developments rendered the idea more practical. Among these was the development of the three-level solid-state maser, which promised an amplifier for ground receivers so much more sensitive than traveling-wave tubes that the transmitting power could be reduced by a factor of 100.[164]

In July 1958, Pierce and Rudolf Kompfner attended a conference on satellite communications that was part of a summer study run by the National Academy of Sciences for the Air Force.[165] They returned to Bell Laboratories advocating an experiment on passive, low-lying spherical satellites as a first step, but they were unable to interest the newly formed Advanced Research Projects Agency in the Department of Defense. The DoD was then pursuing active satellites, including the technologically demanding goal of an active geosynchronous satellite. Instead, Pierce and Kompfner got help from another new agency, the National Aeronautics and Space Agency (NASA). NASA's predecessor, the National Advisory Committee for Aeronautics, had since 1956 been harboring a project to float a high-altitude balloon to measure atmospheric air densities. Pierce and Kompfner proposed that NASA use that balloon as a passive satellite for bouncing electromagnetic messages between Bell Laboratories in New Jersey and other ground stations. Early in 1959, NASA began work on Echo, an aluminized mylar balloon satellite with a diameter of 100 feet. AT&T, meanwhile, began building a major transmitting and receiving station at its Holmdel, New Jersey, laboratory.[166]

The managers of Project Echo at Bell decided to use for the receiver the lowest-noise antenna then available, the "horn-reflector" antenna, which had been invented at Bell and developed in the 1950s for use in overland microwave communications systems. Combined with this would be the traveling-wave maser developed by DeGrasse, Schultz, Dubois, and Scovil. To this time, the maser group's biggest customer for traveling-wave masers had been Bell Laboratories' Whippany facility, which used the masers in military ground-based radar. Now the group began to work with antenna specialists at Holmdel to design an Echo receiver with the lowest noise possible. What was important was the total noise, made up of the individual noise contributions of the antenna, the maser, the circuitry connecting them, and the "sky noise," that is, the background static contributed by the thermal microwave radiation of gases in the atmosphere and the radio waves emanating from cosmic sources. In an elegant feat of engineering, the two groups achieved a total noise temperature of 18.5°Kelvin (−254.7°Centigrade). This was about 3°K more than they had expected on the basis of independent measurements the maser group had conducted on the traveling-wave maser and the antenna group had conducted on the horn-reflector antenna. They concluded that their measurements on these components must somehow have contained errors.[167]

The first successful Echo satellite was flown in August 1960. The receiver at Holmdel contained a modified version of the DeGrasse-Scovil-Ohm-Hogg apparatus. Echo transmitted a tape-recorded message by President Eisenhower from the Jet Propulsion Laboratory to Holmdel on the first day it flew, and later in the month it relayed to Holmdel a transatlantic signal from the French Centre National d'Etudes de Télécommunication. A major market for satellite communications seemed to be in the making. AT&T projected that the volume of international telephone communications would increase fivefold over the coming decade. Television, data facsimile, and telexing would also be important users.[168] AT&T still had a near monopoly on U.S. domestic and international telephone service, but it was not at all clear that the company would be able to dominate satellite communications. This was a technology to which not only the established communications companies, such as AT&T, Western Union, and RCA, but also aerospace companies, such as Ford and Hughes, could bring expertise.[169] Hughes Aircraft, indeed, was already seeking contract money for an innovative active satellite that would function in a geosynchronous orbit. AT&T moved quickly. In October 1960, it proposed to NASA an experimental active satellite, the Telstar, to be placed in a low-altitude orbit. This would be followed in phases by additional satellites, leading to a commercial system for transatlantic service.[170]

Bell Laboratories now became interested in complementing this communications satellite work with radioastronomy. In February 1959, on Kompfner's instigation, the laboratory wrote to Australia seeking an astronomer who would spend one year at Holmdel, doing research "of his [sic] choosing" with the maser–horn antenna apparatus and advising the communications workers "on problems which they expect to encounter in their own undertakings."[171] In March 1961, Kompfner again asserted Bell's "growing interest in techniques which overlap those used in radio astronomy" to A. C. B. Lovell of Great Britain's Jodrell Bank radioastronomy station.[172]

Townes and his students were then in the midst of their experiments at the Naval Research Laboratory telescope. Kompfner, as part of his recruiting activities, invited them to Holmdel to see the horn antenna. As a result of that visit, Arno Penzias joined Bell Laboratories after completing his radioastronomy thesis with Townes in 1961. He settled into a research program that was 70–80% astronomical research and 20–30% radio communications research.[173]

When the first Telstar was placed in orbit in July 1962, Telstar II was already in the planning stage. Penzias's horn reflector was preempted for a time so that a new maser, which amplified at 7.35 centimeters, could be added for the Telstar experiment.

By this time, however, AT&T's attempt to gain the same monopolistic control over transatlantic satellite communications that it had over transatlantic cables was already doomed. The Department of Justice's interest in preventing such a monopoly, presidential concern about the nation's image abroad, congressional qualms over the use of publicly funded research for private profit, and a mosaic of other issues within the government and polity had found expression in the Communications Satellite Act. Passed into law in August, the Comsat Act created a regulated private corporation in which no more than 50% of the voting stock could be held by communications companies. Telstar II, orbited in May 1963, marked the end of AT&T's Telstar program.[174]

Meanwhile Bell Laboratories had hired a second radio astronomer, Robert W. Wilson. With the satellite program reduced, however, the radioastronomy program was cut back to the equivalent of one full-time researcher. Penzias and Wilson each worked at astronomy half-time. They used the horn antenna, enriched with the 7.35-centimeter Telstar maser and with apparatus of their own making.[175] To test this equipment, they turned it to a region of our galaxy just outside the Milky Way. At 7.35 centimeters, they expected no radioemission from the Milky Way's halo. Instead they got an excess signal, in the form of radio noise, corresponding to a temperature of about 3°K. Unlike the Project Echo group before them, Penzias and Wilson had equipment that definitively allowed them to locate the excess noise as coming from or through the antenna, but they could not correlate it with any terrestrial or celestial phenomenon they could think of, nor could they trace it to the antenna itself.

Penzias and Wilson were stumped until, in 1965, they were put in contact with Robert Dicke's group at Princeton University. which was investigating the theoretical implications of a cyclic, "big bang" universe, one that periodically returns to a phase in which all matter and radiation are concentrated in a state of extreme density. What the Princeton group found was that remnants of the high-density radiation from the most recent contraction should still fill the universe and should in fact correspond to the "noise" Penzias and Wilson had detected. There now began a systematic research, by Penzias and Wilson and many others, to establish that the excess radioemissions were indeed coming from this relict radiation. In 1978, Penzias and Wilson were awarded a Nobel prize for their discovery of the cosmic microwave background.[176] Such an intertwining of cosmogony with the development of military radar and the competitive struggle to dominate the nascent satellite communications market is a striking example of how scientific, commercial, and military maser programs fed off each other.

It is clear that the relation between scientific and military applications of the maser does not fit any simple model of "spinoff" in which a fully formed military technology is subsequently adapted to civilian products. Instead, there was a continual bouncing back and forth among R&D for military, commercial, and scientific uses. Several factors drove this three-sided ping pong.[177] One that emerges from the cases outlined here is the procedures of U.S. research institutions, both university-run Department of Defense–oriented laboratories such as Willow Run and Lincoln Laboratory and industrial companies such as AT&T. These organizations had in place arrangements that facilitated the transfer of technology among the three sectors. We have seen at Willow Run and Lincoln Laboratory the liberal use of university consultants, who pursued academic projects at the same time that they advised on government programs. We have also seen, in the case of Weston Vivian of Willow Run, how the fact that university programs and courses were open to laboratory staff could push laboratory projects to be more academic. It is also true that the laboratories made their facilities available to university students, hired some of them (as we shall see below), and provided laboratory scientists as occasional teachers at their universities. At Bell Laboratories, we saw the deliberate cultivation of scientific fields relevant to civilian or military technologies. No doubt further research will turn up many other specific transfer mechanisms.

Science was clearly advanced in this process. The radar exploration of Venus at Lincoln Laboratory[178] and the work of Penzias and Wilson at Bell Telephone Laboratories profited from the ability of scientists to use hardware that was originally developed for military or commercial purposes. But the military and commercial technologies also profited. Research on astronomical microwave receivers had the potential to advance the whole state of the art of low-noise receivers, with payoffs for radar, countermeasures, and satellites. Laboratory administrators knew this when they signed off on scientific projects. It is, of course, precisely the technologies that can easily be transferred among military, scientific, and commercial uses that benefit from the way in which American scientific institutions and American technical workers span all the sectors of R&D: academic, defense, and industrial.[179]

In the event, the microwave maser, for all the elegance of its conception, proved too complicated in the auxiliary systems it demanded for all but a handful of special applications. The U.S. maser industry, as a consequence, never became a large one.[180] But even before this settling out had occurred, and before scientists had had an opportunity to reevalute their initial enthusiasm for the solid-state maser's promise, a new extension of the maser principle took place that would sustain the excitement and utilize the research capacity that had developed around the microwave maser. This was an extension, not to a new class of materials, as in the transition from the ammonia beam to the paramagnetic crystal maser, but to a new spectral region, the infrared and optical.

3

The Laser

The Idea of the Laser

Townes first conceived his ammonia beam maser as a way to produce electromagnetic radiation of about 0.5 millimeter, but by March 1952 he had shifted his immediate goal to generating waves 1.25 centimeters long.[1] The early solid-state masers also produced radiation in the centimeter range: Scovil, Feher, and Siedel's gadolinium ethyl sulfate three-level maser radiated at about 3 centimeters, as did the Kikuchi team's ruby maser. The maser made by McWhorter and Meyer at Lincoln Laboratory operated at about 10 centimeters, and that built by Bloembergen and his collaborators at Harvard had a wavelength of 21 centimeters. Thus, the problem of generating millimeter and submillimeter waves remained open.

Millimeter and submillimeter waves were at the forefront of radio engineering research during the second half of the 1950s. The director of the Stanford University Microwave Laboratory, Edward L. Ginzton, wrote in an editorial in 1956, "Those of us who have specialized in [microwaves] must anticipate either more prosaic engineering applications or a change to some other branch of science. . . . Some will think of exploring the higher regions of frequency lying beyond microwaves—to try to bridge the gap between the radio and infra-red radiation . . . the generation of submillimeter waves, lying . . . between 300 and 3000 kmc [i.e., between 0.1 and 1 millimeter], appears as fascinating and promising today, as the microwave region appeared in 1936."[2]

A University of Illinois scientist, reviewing the field in January 1963 at a talk to the Millimeter and Submillimeter Conference in Orlando, Florida, noted that there had been five prior conferences, in 1951, 1953, 1955, 1959, and 1961. He listed over 150 research and survey articles that had been published during the four years from 1959 through 1962.[3]

The attack on millimeter and submillimeter waves was carried out by a variety of methods. Klystrons were operated down to 5.5 millimeters by the middle 1950s. Magnetrons were made to radiate as low as 2.6 millimeters, and laboratory versions of traveling-wave tubes had by 1957 produced waves as small as 1.5 millimeters.[4] Harmonic generation in a wide variety of nonlinear systems was tried. Researchers were also exploring Cerenkov radiation, in which a dielectric is caused to emit radiation by the passage of a beam of electrons alongside it, and "undulator" devices, in which the radiation is emitted by a beam of electrons forced to travel a sinusoidal path by a nearby sinusoidal but static magnetic field.[5] The maser was considered one of the more promising avenues of approach to submillimeter and millimeter waves.[6] Theoretical proposals for building masers at these wavelengths, and even some experimental projects, came from a range of U.S. research institutions, including universities, industrial laboratories, and Department of Defense–sponsored university laboratories.[7]

From the idea of extending masers into the millimeter and submillimeter region to that of moving still farther up the electromagnetic spectrum, to infrared and even optical radiation, was a reasonable jump. Robert Dicke's work offers an example of the close connection between millimeter and infrared masers. In early February 1956, Dicke jotted down in his idea notebook a trio of related inventions. The purpose of the first was to convert Townes's ammonia beam maser into a millimeter-wave generator (or amplifier) by making use of higher-energy transitions in ammonia or ammonialike molecules; the second was intended to achieve higher output powers by arranging the source of ammonia molecules in a circle around the microwave cavity and altering the focuser appropriately; the third was an invention to use this "circular maser" to generate waves in the region from 0.25 millimeter up to 30 microns (0.03 millimeter), that is, from the far to the intermediate infrared (figure 3.1). To enable his maser to work in this part of the spectrum, Dicke substituted for the microwave cavity a pair of parallel reflecting plates. Such a pair of plates was a well-known feature in optics experiments, where it was known as a Fabry-Perot etalon.[8]

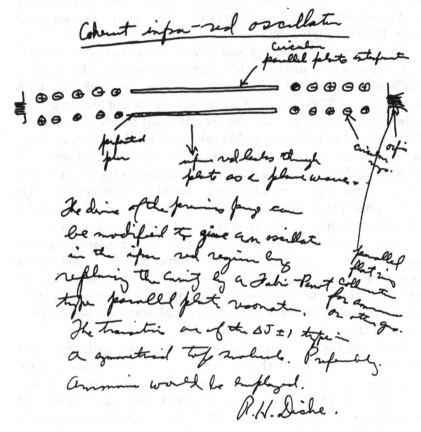

Figure 3.1
Robert H. Dicke's notebook entry for 12 February 1956 shows an idea for a
maser generating radiation in the far and intermediate infrared. (Courtesy R.
H. Dicke)

Other scientists recall having thought about, or discussed, infrared and optical masers in 1956 and 1957.[9] Indeed, William J. Otting of the Air Force Office of Scientific Research had approached senior physicists at the Columbia Radiation Laboratory early in 1957 to suggest that they undertake work in this area.[10] Henry Motz at Stanford University produced light from his undulator experiments, although he treated it as a side result, ancillary to the main job of millimeter generation.[11]

Yet if the idea of moving to infrared and even optical frequencies was not wholly absent for maser researchers, nevertheless the amount of work expended on these frequencies was minuscule through 1957, in comparison with the research done on the 0.1–10 millimeter region. Dicke's ideas, for example, though patented, were not investigated at the RCA Sarnoff Laboratories.[12]

There were good reasons for the lack of attention.[13] First of all, it should be remembered that masers were only one of many approaches to the generation of coherent radiation in the millimeter and submillimeter region. The very existence of a multipronged attack on 0.1–10 millimeter radiation created an environment that fostered work on masers of these wavelengths. Townes's own invention of the ammonia beam maser gives substance to this contention, for his chairmanship of a committee to promote millimeter-wave generation was the proximate stimulus for his maser idea. The programmatic concern with this region of the electromagnetic spectrum led DoD contract monitors to direct the activities of maser scientists toward it. Signal Corps scientist Israel R. Senitzky, for example, visited the Hughes maser group in February 1957. "At present," he reported, "most of the [ir] thinking is in the area of three-energy-level masers. [I] suggested that substantial effort also be expended on the two-energy-level maser, since submillimeter-wave generation and amplification will result only from the latter type."[14] The series of conferences on millimeter waves that were held throughout the 1950s surely provided additional motivation to maser people.

But why was there no similar multifaceted attack on the problem of building infrared and visible coherent generators in the second half of the 1950s? I would point to three factors. First, the technical basis for such an attack was less developed. Second, the need for oscillators in this spectral region seemed less pressing. Third, established habits of thought acted to mitigate interest.

The technical basis must be understood to comprise the available equipment, techniques, and ideas. Klystrons and magnetrons were generating radiation in the millimeter region only with the greatest difficulty, and it was patently obvious that they could not be extended into the infrared. Harmonic generation gave less and less power as one went to shorter and shorter wavelengths. Masers could not be ruled out as infrared generators, but neither could they be counted on, for the amount of power needed to achieve population inversion rose as the cube of the frequency, and there was considerable doubt as to whether this power requirement could be met. In addition, the cavity resonators used in masers needed to be the size of the wavelength they generated, and it would be technically impossible to machine cavities at wavelengths in the infrared. It was the lack of good ideas for overcoming these problems that led Townes to reject Otting's suggestion for a review of methods to extend masers toward the infrared in early 1957.

Second, the need was not as pressing. For spectroscopic applications, where only small amounts of energy are needed, incoherent sources were available. The amount of radiation per unit frequency interval that thermal sources emit increases with the square of the frequency. Hence these sources, which gave so little power in the submillimeter range as to be unusable, had long been acceptable in the visible and infrared.[15] As for military applications, infrared radiation was used mainly for passive sensing of heat emitted by such targets as missiles and tanks. As a result, there was little demand for generators.

It might be objected at this point that since the military was in fact interested in the passive sensing of infrared radiation, it should have wanted sensitive infrared amplifiers. Why, then, was there no drive to develop infrared masers at least as amplifiers, if not as oscillators? The explanation is that the targets the military wished to track put out continuous spectra, whereas maser amplifiers received radiation only in narrow bands of frequencies. In addition, although microwave masers had exceptionally low noise, the noise temperature for masers increases directly with frequency, and so there was no reason to assume that infrared maser receivers would be more sensitive than detectors already in use.[16]

Lack of clear promise, on the one hand, and of felt need, on the other, was supplemented by a third ingredient: habit. There was an unspoken presumption that the electromagnetic spectrum would be conquered step by step. With radio waves, ultra-high-frequency waves, and microwaves out of the way, millimeter and submillimeter waves appeared to be the new frontier. To carry out sustained research on masers in the infrared and visible ranges in this environment required a certain measure of daring and imagination.

At the end of the summer of 1957, Townes began a concerted attack on the problem of an infrared maser.[17] He later wrote, "For some time after the invention of the maser I had been looking forward to pushing its wavelength into the infrared region, though I had been too preoccupied with its other uses to spend very much time in that direction and was simply hoping that perhaps a good idea would strike me about how it could be done well. The failure of any good ideas to occur made me finally, in September of 1957, decide that I should sit down and overtly think through the problem more systematically, and thus plan the best way to make such a maser in the infrared even though the lack of just the right ideas might make such a development clumsy."[18]

Townes wrote down an expression for the population inversion necessary to obtain oscillation in a masing gas. This equation contained the ratio of the frequency of oscillation to the atomic linewidth, as well as a quantity that measured the rate at which the maser cavity leaked energy and a number of other frequency-independent factors that are not of concern here.[19] Looking at this formula (figure 3.2), Townes recognized that under certain circumstances, the ratio of frequency to linewidth should be constant whatever the frequency. Thus, if the cavity loss were not strongly frequency-dependent, the population inversion as a whole would be largely independent of the frequency. "I suddenly realized that maser techniques could just as easily be applied to the visible region and in fact visible waves would probably be easier than the far-infrared, because the equations for an oscillating system showed that no more excited atoms or molecules were necessary for a visible oscillator than for a far-infrared one, and techniques in the visible range were already well developed."[20] In this leap over submillimeter and far infrared into the visible, Townes had had the first of his "good ideas."

Townes now sketched out some preliminary ideas for a visible-wavelength maser. The cavity might be a box with sides about 1 centimeter long—hence enormously larger than the roughly 0.00003-centimeter wave he was considering. Such a large cavity could sustain many modes, but Townes thought that the most

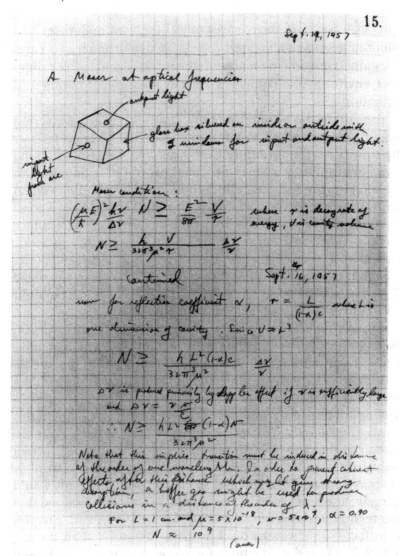

Figure 3.2
Charles Townes committed his early ideas for an optical maser to his Computation Book on September 14, 1957. (Courtesy South Carolina State Museum)

favored mode might suppress the others because of nonlinearities. It might be silvered to reflect at the frequencies of light and provided with holes through which radiation might be fed in or extracted. The working substance might be a gas. Townes initially thought of the vapor of thallium. Excitation could be achieved by optical pumping—that is, by supplying energy through irradiation—using an incoherent light source.[21]

Sometime in October, Townes, who had a consultantship with the Physical Research Group at Bell Telephone Laboratories, discussed his ideas with Arthur L. Schawlow, a member of the technical staff (figure 3.3). The two men were close both professionally and personally. Schawlow had been a Carbide and Carbon Chemicals Corporation postdoctoral fellow under Townes in 1949–1950 and then a research associate in 1950–1951. During that stay, he had met and married Townes's younger sister, Aurelia. Townes and Schawlow had collaborated on papers and on a magisterial monograph, *Microwave Spectroscopy*, published in 1955. Schawlow himself, in 1956 and 1957, had independently given some thought to the problem of making an infrared maser.[22] That October he and Townes decided to collaborate.[23] The work would be done as part of Townes's Bell Laboratories consultancy. It was an appropriate project for Bell, which had both an active program on solid-state masers and their applications and a program on millimeter-wave communications. No management decision had been made to push masers into the optical range; nevertheless, research scientists such as Schawlow were expected to strike off in directions of their own, as long as these studies had potential relevance to communications.

On October 25 and 28, Townes talked to a Columbia University graduate student, R. Gordon Gould, about lamps for optically pumping thallium.[24] Gould was doing his thesis under another member of the Columbia Physics Department, Polykarp Kusch, on the energy levels of excited thallium, and he was using optical pumping to raise the thallium into the state he wanted to observe. These conversations alarmed Gould, he later testified, because he himself had been thinking about how to design an optical maser, or LASER, as he called it (for Light Amplification by Stimulated Emission of Radiation). He therefore noted down his own ideas and, on November 13, had them notarized.[25] Gould proposed to contain the working substance in a meter-long tube terminated at

Figure 3.3
Arthur L. Schawlow adjusts a ruby laser during a 1960 experiment at Bell
Laboratories. (Courtesy AIP Center for History of Physics)

the ends by reflecting plates (see figure 3.4). Hence he began, as Dicke had, with a Fabry-Perot etalon, but a long and skinny one as against Dicke's pancake-shaped resonator. Gould next derived the condition for laser oscillation, arriving at a formula that gave "the usual condition for maser oscillation." He then mentioned as a possible form of excitation the optical pumping he and Townes had discussed, and as a possible medium an alkali vapor such as potassium vapor. The last two of his eight pages of notes were devoted to possible applications, which included spectrometry, interferometry, photochemistry, light amplification, radar, and communications. "Perhaps the most interesting and exciting possibility," Gould wrote, "lies in focussing the beam into a small volume . . . with a tremendous factor of energy concentration. A solid or liquid placed at that focal point would be heated at the rate of [about] 10^{16} °K/sec. If the substance were heavy water, nuclear fusion temperatures could possibly be reached before the particles were dissipated."

Although he was working on a PhD thesis on molecular beams, Gould later recalled that, at the time, he thought of himself primarily as an inventor. He had received a BA in physics from Union College in 1941, and an MS in optical spectroscopy from Yale University in 1943, but after he had finished his wartime military service, he had chosen to take part-time jobs to support himself while he invented. Gould had designed an improved contact lens, and he had tried to make synthetic diamond. He had concluded, however, that he needed a deeper scientific background for the kinds of inventions he contemplated, and he had begun to take graduate courses at Columbia University in 1949. In 1951, he had embarked on a PhD, and in 1954 he had enrolled as a full-time PhD student. In November 1957, he still had his thesis to write. But the laser's potential for applications, he later recalled, made a strong appeal to his inventor's heart.[26]

Townes and Schawlow, for their part, continued to refine their ideas through the fall and winter of 1957–1958. Schawlow made a survey of possible maser materials, and he and Townes decided that thallium would not work well and that potassium vapor had a number of advantages, including the easy availability of equipment to measure its lines. Schawlow, prodded by a colleague at Bell Laboratories, gave special attention to the problem of mode selection. He suggested that if a laser medium were placed in a cavity with reflecting end walls whose diameter was small with

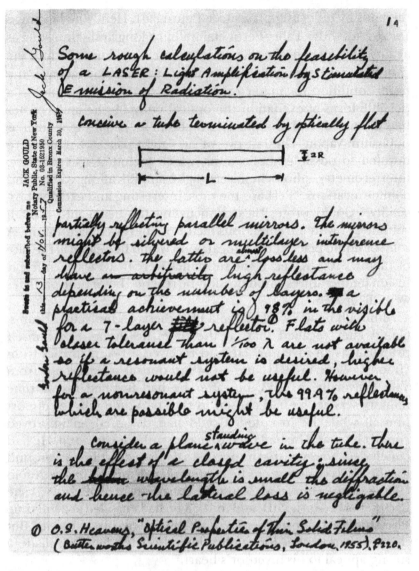

Figure 3.4
The first page of the notes that Gordon Gould had notarized on November 13, 1957. (Courtesy R. G. Gould)

respect to the cavity length, and with transparent side walls, only the modes traveling nearly parallel to the cavity axis would oscillate.[27] This was, again, a structure related to the Fabry-Perot etalon, as both men recognized. Schawlow and Townes proposed to strengthen mode selection by using the fact that, because of the small ratio of the wavelength to the cavity dimension, the light from different modes, exiting from the ends, would form distinguishable diffraction patterns. The light from the desired modes could then be separated out from the other light by means of a lens.

A central problem Townes and Schawlow had to face was whether it would be possible to supply sufficient power to an active medium. They tackled this issue, first, in the course of a general discussion of the conditions for infrared and optical masers. For a medium made up of gas atoms, of atomic weight 100, and with two energy levels separated by an energy corresponding to a 1-micron wavelength, they calculated that the minimum number of atoms needed in the upper level, in order to have enough energy released through stimulated emission to balance energy losses, was 5×10^8 atoms/centimeter^3. The minimum power to raise that many atoms through a 1-micron transition would be 0.8×10^{-3} watt. "Supply of this much power in a spectral line does not seem to be extremely difficult."[28] Schawlow and Townes returned to the question of power later in their paper in the context of a discussion of a specific possible maser: potassium vapor housed in a resonator 10 centimeters long and 1 square centimeter in cross section and pumped by light from a potassium lamp. They calculated the minimum power for this case to be 1.2×10^{-3} watt, a value that was "quite attainable." They had already increased the power output of a small commercial potassium lamp, rated at 0.08×10^{-3} watt, to 0.6×10^{-3} watt, half of what was required.[29]

In August, after Bell Laboratories had submitted a patent application, Schawlow and Townes sent their manuscript to the editors of the *Physical Review*. This was the leading publication outlet for American physicists, and its appearance there would ensure that their work would have a wide audience among physicists both in the United States and abroad.

Meanwhile, Gordon Gould had left Columbia University in March 1958, without finishing his thesis, and had taken a job with TRG, Inc. TRG was one of those American companies that came into being during the Cold War and was sustained by military spending for advanced weapons technology. It was founded as Technical Research Group in 1953 by three men with recent doctorates in the fields of electronics, physics, and applied mathematics, and at first it functioned as a consulting firm. In 1955, the firm added laboratories and shops. The early contracts were with Department of Defense agencies and contractors, such as the Office of Naval Research, the Air Force Cambridge Research Center, and the Johns Hopkins Applied Physics Laboratory. "At that time, by and large," recalled Lawrence A. Goldmuntz, one of the founders and a former president, "the government was really interested in pursuing things, so if you had a good idea, you'd go down to a government agency and they would generally support it."[30]

TRG worked on antennas and radar, nuclear reactor physics, missile guidance, and the hydrodynamics of naval hull designs. They also had projects in quantum resonance technologies, including a small maser contract, and a program on atomic frequency standards. TRG was an informal and creative company of about 100 employees in 1958, perhaps 30 or 40 of them scientists, and it had ambitions to field its own commercial products. At that time, however, most of its work was on short-term contracts—*Aviation Week* called it a "scientific job-shop"[31]—and 90% of its 1957 funding was for aircraft weapons systems. Gould was hired to work on atomic resonance frequency standards, but he had negotiated an employment agreement that specifically withheld from the company rights to inventions he had made before March 1958. He intended this to cover his laser interests. Since Gould had not yet written up his thesis results, TRG granted him free time in July 1958 for his own work. Gould used that time to work not on his thesis but on the laser.[32]

Gould's 1958 work on the laser was distinguished by its fecundity.[33] He discussed both nonresonant cavities, which he believed would be useful as a test bed for maser materials, for setting frequency standards, for spectroscopy, and for distance measurements, and resonant cavities incorporating Fabry-Perot structures, with uses in communications, radar, and heat processing. He explored a large variety of laser media and methods of excitation, including optical pumping, excitation by impact with energetic electrons in a gas discharge, and excitation through the transfer of energy from an excited "sensitizing" atom or molecule to an atom of the lasing substance ("collisions of the second kind"). As for laser media, Gould made calculations for sodium, sodium with mercury as sensitizing agent, and helium. He discussed, among other possibilities, zinc and thallium vapors excited by transfer of energy from krypton and xenon, molecular iodine pumped by sodium light, and optically excited europium sulfate in water solution. He supplemented these discussions with an enumeration in tables of other atoms that might be led to lase. He talked about parallel mirrors, curved mirrors, and 90° prisms as reflecting elements for the Fabry-Perot. Many of his analyses were, properly speaking, merely suggestive, with the physics not thoroughly worked through. There were many errors. Nevertheless, for sheer inventiveness, Gould's 1958 laser writings were a tour de force.

Gould filed a patent application in April 1959. When the Schawlow and Townes patent on behalf of Bell Laboratories, filed in July 1958, was granted in March 1960, Gould and TRG brought a challenge against it before the U.S. Customs and Patent Appeals Court, on the grounds that although Gould had filed later, he had conceived of the invention first.[34] Since Townes and Gould had both been at Columbia University and had had direct interactions, questions were naturally raised as to whether one of them might have appropriated the ideas of the other.[35]

Conscious, unacknowledged, and unscrupulous borrowings are a fact of scientific life. So are situations in which the same idea occurs independently to several people. There are, moreover, a whole range of cases that lie between these two extremes. Ideas only take root in prepared minds, and it is not always easy either for the outsider or the inventor to separate the preparation from the new seed. How can we make historical sense of this controversy? Let us begin by considering the problem of invention more generally.

It is a common opinion among historians of science and technology that it is usually a mistake to try to link an invention or a scientific discovery to a single individual or instant in time. In a recent study of early radio, Amherst College professor Hugh G. J. Aitken wrote: "We are inclined to think of invention as an act rather than a process because of the bias built into our patent laws. If property rights in a new discovery are to be secured, it is important to be able to establish priority in time. ... This bias, however, should not be allowed to corrupt our historical interpretations ... invention [is] a process with considerable duration in time, one to which many individuals contribute in a substantial way."[36]

Thomas S. Kuhn, in an article entitled "Energy Conservation as an Example of Simultaneous Discovery," pointed out that for most of the period from 1830 to 1850, no two of the twelve scientists who enunciated forms of the concept of energy conservation were saying substantially the same thing. "What we see . . . is . . . rather . . . the rapid and often disorderly emergence of the experimental and conceptual elements from which [the theory of energy conservation] was shortly to be compounded."[37] In a 1962 article, Kuhn built on this analysis and identified the existence of a "structure" within the history of a scientific discovery. There are, he suggested, inherent uncertainties in the date of a discovery and the identity of the person to be associated with it that can be analogized to the uncertainties that quantum mechanics decrees for the position and momentum of a physical particle.[38]

In the specific case of Townes, Schawlow, and Gould, I have no evidentiary base for judging whether or not Gould had already been thinking about the laser before Townes summoned him to discuss thallium lamps for optical pumping. But once Gould had the idea, then I believe that the documents leave no doubt that he developed it in a unique way. The differences between his work and that of Townes and Schawlow had to do with the differences in background, style, and knowledge that obtained among the three men.

Gould's master's degree at Yale had been in optics, whereas Townes was a microwave physicist. It is therefore not surprising that Townes's first sketch of a resonant cavity for an optical maser, in September 1957, resembled a microwave cavity, whereas Gould's sketch, in November 1957, was of an optical element, a Fabry-Perot etalon. Gould was primarily an inventor, whereas Schawlow and Townes were primarily physicists. Moreover, the three were working within very different literary forms. Gould was writing first an idea notebook, and later a proposal for a contract, while Schawlow and Townes were writing an article for a refereed journal. Therefore it is not surprising that Gould sent up a fireworks of partially developed suggestions, whereas Schawlow and Townes published a general discussion of the physics of optical masers, followed by a carefully worked through particular case. Only at the end of their article did they permit themselves a few suggestions, notably for a helium-pumped cesium laser, and a comment on the possibility of solid-state lasers. Townes and Schawlow were by far the more accomplished and knowledgeable physicists. This is reflected in the thoroughness of analysis in their paper as compared with Gould's notebook.

Like the work of Kuhn's twelve pioneers in energy conservation, the contributions of the Schawlow-Townes collaboration and of Gould in 1957 and 1958 were not entirely coextensive. The sum of the two was greater than either one. It is important to note, however, that that sum was still not big enough.[39] This is proven by the fact that Townes's group was not able to build a potassium laser and was, in fact, to abandon potassium for another working substance, while TRG was to succeed in reducing one of Gould's suggestions to practice only in 1962, two years after the first laser had been operated. To build working lasers by the year 1960, the contributions of others would be needed. That is to say, the uncertainty in inventors would turn out to be larger than the trio of Schawlow, Townes, and Gould.

The Construction of the Laser

The laser proposals could scarcely have been put before the U.S. scientific community at a more auspicious time than in the year after Sputnik. In 1958 and 1959, as we saw in chapter 1, the nation offered unprecedented admiration, resources, and freedom of action to scientists. In this balmy climate, teams at half a dozen laboratories—Columbia University, Bell Telephone Laboratories, TRG, IBM, Hughes, and American Optical Company—all entered the race to construct the first operating laser.

Townes and Schawlow, after they had finished their joint paper, had agreed between themselves that Townes should pursue the potassium laser.[40] Townes was attracted to it, not only because the numbers showed it should work, but because potassium was a simple monatomic gas whose properties were known, and he wanted a system that was susceptible to thorough analysis. "My style of physics," Townes has said, "has always been to think through a problem theoretically, analyze it, and then do an experiment which *has* to work. If it doesn't work at first, you make it work. You analyze and duplicate the theoretical conditions in the laboratory until you beat the problem into submission."[41] His preliminary calculations showed Townes that a potassium laser would have a highly monochromatic output. It would have drawbacks: low efficiency (about

0.1%), and a power output of fractions of a milliwatt. It would, however, be a first step toward devices that Townes envisaged would be useful for high-resolution spectroscopy, improved interferometry, space-based communication, and other civilian and military applications.[42]

Townes applied to the Air Force Office of Scientific Research for funds to initiate work on a potassium laser at the Columbia Radiation Laboratory. OSR support came quickly. As we have seen, its Physics Division was already trying to stimulate work on higher-frequency masers. William Otting of the OSR later wrote of the time between the submission of Townes's proposal in July 1958 and the authorization of the contract in September, "I think this was the shortest time between receipt of a proposal and the issuance of a contract in the history of AFOSR." Townes recruited two graduate students, Herman Z. Cummins and Isaac D. Abella, to work on the project.[43]

While Townes explored the potassium laser, Schawlow at Bell Laboratories elected to work on solid-state lasers. He began with ruby, because of its broad absorption bands, because it was in plentiful supply from the ongoing maser program, and because it had some interesting and unexplained lines in its spectrum.[44]

During the period when he was negotiating his Air Force contract, Townes was also asked by the Office of Naval Research to organize a conference on the technological uses of quantum resonances. These techniques were revolutionizing microwave technology, and the ONR scientists believed that an international meeting could spur further progress. Townes put together an organizing committee representing some of the chief American players in the maser field: George Birnbaum from the Atomic Physics Section of Hughes Research Laboratories, Nicolaas Bloembergen from Harvard, Robert Dicke from Princeton, Malcolm Strandberg from MIT, Benjamin Lax from Lincoln Laboratory, Rudolf Kompfner from Bell Telephone Laboratories, Anthony Siegman from the Stanford University Electronic Laboratories, G. J. Stanley from California Institute of Technology, and Charles Kittel from the University of California at Berkeley. They scheduled the conference for September 1959, and the name they chose for their meeting, "Quantum Electronics—Resonance Phenomena," underlined the confluence of engineering and physics that lay at the heart of the field.[45]

Two other scientists at Bell Laboratories joined Schawlow in working on lasers. One was Ali Javan, who like Schawlow was employed in Sidney Millman's Physical Research Section. The other was John H. Sanders, who was in Kompfner's Electronics and Radio Research Section.

Javan had completed his PhD under Townes in 1954 and had stayed on in Townes's group for four years as a postdoctoral fellow, working in microwave spectroscopy and masers (figure 3.5). He had first heard about the optical maser research from Schawlow in late April 1958, when he had talked with him while interviewing for a job with Bell Laboratories.[46] Javan joined Bell in August 1958, and by October he had started systematic studies in preparation for laser research. He decided to use gas as his working medium because he believed that the simplicity of gases made them better vehicles for the study of physical processes.[47] But Javan thought that Schawlow and Townes's scheme of optical pumping would not

Figure 3.5
Ali Javan at Columbia University with an early beam apparatus. (Courtesy Smithsonian Institution)

supply enough power. Two other approaches looked more promising: direct electron excitation, with pure neon as the medium, and collisions of the second kind. In the latter scheme, the discharge tube would be filled with two gases chosen in such a way that atoms of the first, excited by the impact of electrons, could transfer their energy to atoms of the second, leaving the latter in an upper laser level. A number of gas combinations had a structure of energy levels that satisfied these requirements; among these, Javan preferred the combination of helium and neon, which would allow him to employ techniques recently discovered for studying energy transfer between noble gas atoms (figure 3.6).[48]

Meanwhile John Sanders, an experimental physicist from Oxford University, had been invited by Kompfner for the months from January to September 1959. Kompfner was then deeply engaged in satellite communications research, but he was excited about optical communications, which he saw as the likely next step in the technology. In October 1958, he had sent Sanders a preprint of the Schawlow-Townes paper and suggested that Sanders join the "attempts to push the 'Maser' towards the infra-red." Sanders enthusiastically accepted.[49] With less than a year of research time before him, he elected a cut-and-try attack. He decided to excite pure helium in a discharge tube inside a Fabry-Perot, and to search for lasing by varying the discharge parameters. The maximum length over which a pair of Fabry-Perot plates could be aligned with the incoherent sources then available was about 15 centimeters. Sanders was therefore restricted to a cavity of this length or shorter.

Javan, however, saw Sanders's restriction on length as a crucial flaw.[50] Javan expected the gain of a gas laser to be so low that Sanders's cavity would be short by a considerable factor. Since it would not be possible to align a Fabry-Perot etalon of the length he wanted, Javan chose the alternate strategy of first determining the parameters for adequate gain. This was to be followed by placing the optimized gas mixture within a long Fabry-Perot cavity and then, as a final step, adjusting the end plates until the start of lasing revealed that they were in alignment. Javan worked in close consultation with William R. Bennett jr., a spectroscopist at Yale University. Bennett was, like Javan, enthusiastic about the prospects both of elucidating the physical processes going on in the discharge and of achieving a laser.

At TRG, in September 1958, president Lawrence Goldmuntz pressed Gould to reveal the inventions, still unspecified, that he

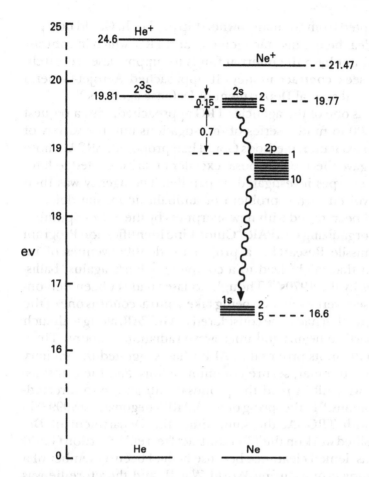

Figure 3.6
Excitation of neon by collisions of the second kind. The energy difference
between the 2³S level of helium and the 2s levels of neon is small enough to
permit appreciable transfer of energy, while that between the 2³S level of
helium and the 2p levels of neon is too large to excite neon to these states.
When the totality of other excitation and relaxation processes was taken into
account, Javan and Bennett were able to show that a population inversion
could subsist between the higher-energy 2s states of neon and the lower-
energy 2p states.

had exempted from company ownership in March. Gould now put his laser ideas before his management, and TRG, which in contrast to Bell did not have the internal funds to support laser research, began to seek contract money. It approached Aerojet-General Company and several Department of Defense agencies.[51]

ARPA was one of the agencies TRG approached, with a request for $300,000 to fund a series of investigations into the variety of media and excitation methods Gould had proposed. ARPA, whose reviewers gave the TRG proposal excellent marks, wanted to have all the laser types investigated in parallel. The agency was then deeply involved in the problem of antiballistic missile defense, which had been raised with new sharpness by the Soviet Sputniks. It was just organizing GLIPAR, "Guide Line Identification Program for Anti-missile Research," a project to identify avenues of investigation that might lead to a complete defense against ballistic missiles by the 1970s.[52] Though no laser had yet been demonstrated, lasers were even then being taken into account as one of the technologies that had to be considered by GLIPAR, along with such ideas as particle beams and microwave radiation weapons. Thus, whereas TRG in its proposal to ARPA had suggested the military applications of radar, secure communications, and the transmission of power, ARPA read the proposal with an eye to directed-energy weapons.[53] In the spring of 1959, ARPA negotiated a $999,000 contract with TRG. At the same time, the Department of Defense classified work on the TRG contract "secret."[54] Gordon Gould himself was denied clearance because he had been a member of a Marxist study group during World War II, and therefore he was unable to lead the project, read the reports, or participate directly in the experimental work. He assumed instead the role of an internal consultant, and guru, to the TRG team.[55]

A session on laser research was included in the June 1959 Conference on Optical Pumping at the University of Michigan. Gould presented some unclassified, theoretical calculations arising out of the secret project at TRG. John Sanders reported on his unsuccessful attempts to create a population inversion in helium, contained in a tube within a 7.5-centimeter Fabry-Perot cavity and excited by an electric discharge. Schawlow talked informally about his investigations into the spectrum of ruby.[56]

Irwin Wieder of the Westinghouse Research Laboratories gave a paper arising out of his program to extend the use of optical pumping from gases to solid-state masers. A key problem with optical pumping in 1959 was that the available sources of pumping light were incoherent and gave out light with a wide spectrum of frequencies. Thus only a small fraction of the total light output was available in any particular small spectral range. To aggravate the difficulty, Wieder wanted to pump light into a prominent but very narrow line of ruby, the so-called R line.[57]

Wieder and his collaborator had attacked this problem by making a pumping light out of another ruby crystal (figure 3.7). They used a tungsten lamp to pump the ruby lamp, sending the light into a pair of very broad absorption lines located at a higher energy than the upper level of the R line. Because the lines were broad, the absorption of energy from the tungsten lamp was relatively efficient. The absorbed energy was then transferred to the R level by a process that Wieder estimated to be about 1% efficient. (That is, 1% of the energy pumped into the broad bands showed up as light

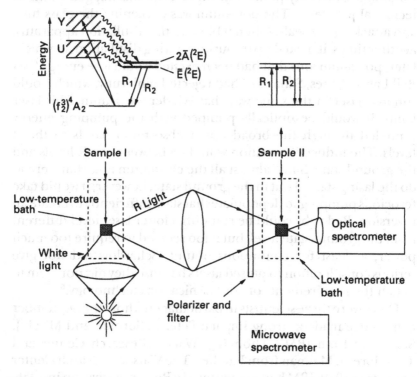

Figure 3.7
Optical pumping of one ruby crystal by another. (Adapted from I. Weider, "Maser Studies," Westinghouse Research Laboratories Quarterly Technical Report 8, February 15 to May 15, 1959)

emitted from the R band.) The ruby then dropped from the R levels back to the ground level, radiating its red R line into the ruby maser. In this way, Wieder could increase by a factor of ten the radiation his maser absorbed.[58]

Now came the September 1959 quantum electronics conference. The timing was propitious. Although the conference had been inspired by microwave developments, and most of the formal papers were on microwaves, much of the informal discussion centered on lasers, and the conference's most important role was to hasten their birth. First, it produced a consensus that lasers were one of the most promising new research areas. As a result, individual scientists began to rearrange their priorities.[59] Second, it offered an arena in which scientists interested in lasers could survey their competition.

And in fact, some of the competitors were having difficulties. The Columbia University potassium vapor laser was bogged down in technical problems. The potassium was darkening the glass tube and attacking the seals. Often it blew up the distillation apparatus; at other times, it picked up impurities during distillation, to which, later, potassium ions that had been excited would lose energy.[60] At Bell Laboratories, Schawlow had rejected pink ruby, which would run as a laser in the same way that Wieder was running his laser lamp. It would be optically pumped with the pumping energy funneled through the broad upper absorption bands to the R levels. The induced transition would be between the R levels and the ground state. Since almost all the chromium ions that were to do the lasing started out in the ground state, however, it would take ferocious pumping to depopulate that state sufficiently to create an inversion. Red ruby could be run with a lower laser level different from the ground-state level, but it too seemed to require too much power, at least for continuous output. Schawlow did not give serious consideration to pulsed lasers because they did not seem to match the requirements of communications technology.[61]

There were three scientists in particular on whom the September conference made a strong impact: Peter P. Sorokin and Mirek J. Stevenson from IBM's Thomas J. Watson Research Center and Theodore H. Maiman from Hughes. The Watson Research Center was new in 1959. IBM had inaugurated a Research Division in 1956, had hired Emmanuel Piore, former chief scientist at the Office of Naval Research, to head it, and had given him the resources to create a "world-class" facility. The Watson Research Center was typical of the laboratories of the decade in the amenities it provided its scientists: a campuslike atmosphere in the pleasant countryside, but close to the cultural resources of a big city, in this case New York.[62]

Even before the quantum electronics conference, the head of the physics section, William V. Smith, had been inspired by the Schawlow-Townes paper to steer his microwave spectroscopy group—which included Sorokin and Stevenson—toward lasers. Smith was convinced that coherent generation and amplification of light were bound to be useful to IBM. He also believed that laser research, as an area of high visibility, could help establish the new laboratory's reputation.[63]

Sorokin had already emerged as the IBM group's chief strategist. He believed that adequate pumping power was the problem that most needed to be addressed. Schawlow and Townes's paper had taught that the longer the light emitted by the excited atoms could be retained within the resonator, the lower would be the minimum required pumping power. But it was not clear how good a reflectivity, and therefore how long a retention time, the mirrors terminating a Fabry-Perot resonator could provide. Sorokin therefore conceived the idea of eliminating the mirror by using total internal reflection to provide the feedback. Then Smith contributed the suggestion that if the crystal chosen had the proper refractive index, total internal reflection could also provide mode selection. Sorokin searched the literature and found that calcium fluoride had the right index of refraction to be the host crystal. He then turned his attention to the ions that, introduced into the crystal as dopants, would constitute the active lasing systems. He sought ions whose spectra included broad bands that were strongly absorbing, and from which energy absorbed could be subsequently channeled down into the upper laser level. In this respect, they would behave like the chromium ions in ruby. However, unlike pink ruby, for which the lower laser level would be the ground state, Sorokin looked for systems in which the lower laser level was normally unoccupied ("four-level systems"). Trivalent uranium and divalent samarium satisfied these conditions.[64]

The effect of the conference was to accelerate IBM's effort. Sorokin recently recalled that "we came back from that conference, and we really decided at that point to drop what we were doing ...

and try to focus on doing some experiment related to this new field that looked like it was going to break."[65]

Theodore Maiman was a member of Harold Lyons's maser group at Hughes Laboratories. Maiman, who had just designed an innovative lightweight ruby maser for the Signal Corps, was at the quantum electronics conference to give a report on physical properties of ruby masers.[66] He was also, by this time, actively studying the production of "hyperfrequency" masers, that is, masers in the millimeter and infrared region, excited by optical pumping and fashioned of salts doped with gadolinium ions.[67]

Maiman mistrusted the emphasis being placed at the time of the conference on gas experiments. He noted that potassium vapor was corrosive and difficult to work with, and that even inert gases such as helium and neon embroiled the experimenter in the complications of vacuum systems and in worries about contaminants. Solids struck him as far more promising. They could give higher power, be more rugged, operate under less restrictive temperature conditions, and make possible smaller-sized devices. Maiman had an excellent appreciation of the importance those qualities could have for practical applications. He was attracted by ruby and was skeptical of Schawlow's dictum that it would be impossible to empty the ground state of pink ruby of half its population. Irwin Wieder had, as we have seen, gotten a figure of about 1% for the efficiency with which energy pumped into the absorption bands would be transferred to the R lines. Maiman took the opportunity to ask Wieder about the details of his experimental work during the conference, and he formed a suspicion that it would be possible to better this figure.

When Maiman returned to Hughes, he threw himself into laser research.[68] He first made a detailed calculation showing that if Wieder's value of 1% efficiency were to prove incorrect, a pink ruby laser would be possible. He therefore decided to work with ruby. If it proved unsuitable, the elucidation of its properties would at least make it easier to choose another, more favorable solid.[69] The project Maiman started was a small one, limited to himself and his research assistant, Irnee J. D'Haenens, and it was funded on the modest level appropriate to exploratory research, out of company funds. He nevertheless felt himself to be under management pressure. He believed that both his immediate supervisor, George Birnbaum, head of the Quantum Physics Section, and his department head,

Harold Lyons, had been convinced by the work of Wieder and Schawlow that ruby held no promise as a laser material.

Maiman adopted a twofold strategy. First, he would find out why the fluorescent efficiency that Wieder had reported was so low. To do this, he needed to measure the strength of the transitions, both radiative and nonradiative, from the bands that would absorb the pumping light to the ground state. He soon found that the strength of the radiative transition was so small as to escape detection. Thus there was no reason to think that pump photons were being wasted through direct transitions to the ground state. Maiman now doubted Wieder's result sufficiently to make a direct measurement of the efficiency. It took a month's work to satisfy himself that for his ruby samples, the efficiency was at least 70%; a series of follow-up experiments convinced him that it was near 100%.

The second element of Maiman's attack was a rethinking of the whole question of pump sources. He first determined that the property of pump lamps crucial to his needs was the brightness. The brightest continuous commercial lamp available was the AH6 mercury arc; if its light were made, with the aid of reflectors, to traverse a ruby several times, he might expect marginal success. He reasoned, however, that this was not good enough, since experimental results were likely to show difficulties not revealed by calculations. Maiman decided to go to pulsed sources and hit on the idea of using photographic flash lamps. He scoured the catalogs for the brightest flash lamps sold and came up with three from General Electric; their helical shape then suggested to him that he could get the greatest density of light into ruby by placing a cylinder of the crystal inside the lamp (figure 3.8).[70]

Maiman's research, carried out in the fall and winter of 1959–1960, was not much noted by the other contestants in the Laser Derby. Maiman was tucked away on the West Coast, he was secretive (as, indeed, were some of the other teams), and he was working on a material universally regarded as unpromising. The other entrants continued their work. Starting in September 1959, Townes had taken a two-year leave of absence from Columbia to serve as vice president and director of research at the Institute for Defense Analyses, an agency set up to advise the Advanced Research Projects Agency. In his absence, he had brought in the British scientist Oliver S. Heavens, an expert in the physics underlying highly reflecting mirrors, to help his graduate students with the laser

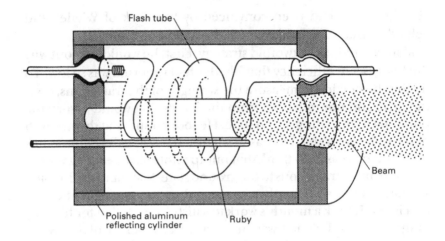

Figure 3.8
Maiman's ruby laser. (Adapted from T. H. Maiman et al., "Stimulated Optical
Emission in Fluorescent Solids. II. Spectroscopy and Stimulated Emission in
Ruby," *Physical Review* 123 (1961), 1154)

project. The Columbia University group had also shifted its emphasis from potassium to cesium vapor, with the optical pumping
provided by helium lamps.

In the spring of 1959, Javan had asked Donald R. Herriott, a Bell
specialist in optical apparatus, to collaborate in the helium-neon
project on a part-time basis. Herriott had started on the optical
problems associated with the project. This involved, first of all, work
on the transparent windows that close off the discharge tube. These
had to be of good optical quality so as not to distort the output
beam. Then mirrors had to be created by selecting optical flats,
testing their flatness, and overlaying them with highly reflective
coatings. A structure had to be designed that would allow the
mirrors to be aligned in parallel, and equipment for detecting the
expected output radiation had to be readied (figure 3.9).

Javan had also arranged to have William Bennett leave Yale and
join Bell Laboratories. Bennett had arrived in September 1959, and
he and Javan started an intensive and meticulous program of
calculating and measuring the spectroscopic properties of helium-
neon mixtures under varying conditions, so as to determine the
factors governing population inversion. These factors comprised
all the excitation processes that tended to populate the upper and

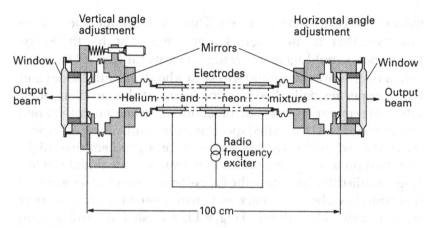

Figure 3.9
The helium-neon laser.

lower laser levels, all the relaxation processes that tended to depopulate them, and the way these processes changed with changes in the parameters of the discharge. Javan and Bennett found that the achievable population inversions appeared to be very small, corresponding to gains in light intensity of the order of 1.5%. This emphasized the need for minimizing losses of energy in the cavity structure and thus underlined the importance of getting the highest possible reflectance out of the mirrors.[71]

Another loss that could be troublesome was the "diffraction" loss due to the finite diameter of the light beam traversing the laser tube. Following Schawlow and Townes's treatment, Javan was approximating this light beam by a plane wave. Because of diffraction loss, even a plane wave traveling exactly parallel to the axis of the discharge tube and exactly perpendicular to the mirrors would lose some energy by leakage around the edges of the mirrors.[72]

In the spring of 1960, two Bell Laboratories scientists, A. Gardner Fox and Tingye Li, showed that the diffraction loss would be less serious than Javan had thought. They obtained this result in the course of an attack on a still more fundamental problem, that of whether the Fabry-Perot etalon can in fact sustain modes, that is, electromagnetic patterns that retain their shape while bouncing back and forth between the end mirrors over long periods of time. Modes were known to exist in closed microwave cavities, but it was not a foregone conclusion that they would persist when the side walls of the cavity were removed. This had led some scientists to doubt whether Schawlow and Townes were right that the Fabry-Perot etalon could serve as a resonant cavity for lasers.[73] Gardner Fox, a microwave waveguide specialist who dealt with modes and their properties daily, had become aware of this problem at an informal seminar on lasers that met once or twice a month at Bell Laboratories and that united spectroscopists, optical scientists, and microwave scientists. He saw that this was a problem that might be solved on a computer: One could assume some initial electromagnetic distribution within the Fabry-Perot etalon and see whether it persisted as it bounced back and forth between the mirrors over many transits. He enlisted Tingye Li, a fresh PhD with a good knowledge of programming, and in February 1960 the two started to simulate the behavior of waves reflected between two mirrors on a computer. By spring, they had obtained the result that modes did exist in a Fabry-Perot cavity. They also had the result that, because the light beam in fact deviated from the model of the perfect plane wave, the diffraction loss would be lower than Javan, Bennett, and Herriott were anticipating.[74] This was a distinct encouragement to Javan and his collaborators as they struggled with a project that seemed to have only a 50-50 chance of success.

In 1959 and 1960, the American Optical Company also entered laser research, through the work of one of its scientists, Elias Snitzer. American Optical's original focus was optical instruments and ophthalmic products, and it had strong capabilities in glassmaking and glassworking. During the 1950s, the company decided to expand its product lines, and it therefore initiated research projects in such new areas as military electro-optics and fiber optics. Elias Snitzer was hired into the research group in early 1959, and he began his tenure with research on the propagation of electromagnetic modes through optical fibers. For the company, this work would strengthen its already considerable patent position in fiber optics and bolster its image in the area in the general scientific community.[75]

There were connections between this optical fiber research and laser work, as Snitzer saw. Since a glass fiber could sustain electromagnetic modes, it could be converted into a laser resonator if one placed mirrors at its ends. This prospect was the more interesting because of the doubts within the scientific community about whether the kind of Fabry-Perot resonator that was being used with gaseous media would work. The glass itself could be made into a lasing material by doping it with a paramagnetic substance such as samarium or ytterbium. The paramagnetic ions could be excited by incoherent light sent in through the sides or the end. Snitzer reasoned that he could concentrate even more pumping light (delivered through the fiber's ends) into the fibers by cladding them with a thicker layer of a glass with a slightly different index of refraction. In early 1960, with two more junior colleagues, Snitzer began to examine a succession of clad glass fibers doped with ions that had fluorescent lines in the visible spectrum. Glass was an unusual choice. All other laser work at that time was being done with gases (potassium vapor, cesium vapor, noble gases) or crystal (ruby, calcium fluoride).[76]

In mid-May 1960, Maiman completed his calculations and preliminary experiments and put together an experimental laser. He used a cylinder of ruby 1 centimeter in diameter and 2 centimeters long, silvered on the ends to make it into its own Fabry-Perot resonator. He and D'Haenens got immediate signs of lasing, but they did not get the abrupt onset of lasing behavior—the appearance of a "threshold"—that they had expected. The ruby, however, was left over from microwave maser experiments and had poor optical qualities. Maiman now ordered special rubies from a leading crystal-growing firm.[77] By this time, however, he keenly felt the pressure to publish. His management, fearful that Columbia University would scoop Hughes, was urging him to make his results known. He himself had just published his finding of the high quantum efficiency of ruby in the *Physical Review Letters*, and he was apprehensive that this would put others on the trail of the pink ruby laser.[78] On June 24, Maiman submitted a brief article, "Optical Maser Action in Ruby," to Samuel Goudsmit, the editor of the *Physical Review Letters*. To Maiman's dismay, Goudsmit turned it down.[79]

Maiman now sent an abbreviated version of his letter to the British journal *Nature*.[80] In addition, Hughes Aircraft Company held a press conference on July 7 to announce the discovery. The event was so exciting to the technical world that the magazine *British Communications and Electronics* picked up the preprint of an article Maiman had submitted to the *Journal of Applied Physics* and, to the young scientist's embarrassment, ran it without permission in their own pages. Moreover, portions of the U.S. scientific community were greeting Maiman's announcement with skepticism.[81] Maiman, though his results strongly indicated lasing, had not dotted every i. The preconception that ruby would not lase was probably one obstacle to acceptance, and the expectation that one of the East Coast laboratories would come in first may have been another. In August, a group of Bell scientists put together a near-reproduction of one of Maiman's set-ups and showed beyond a doubt that the ruby was lasing. But their publication, which came out in the October issue of the *Physical Review Letters*,[82] caused still more confusion. Faced with Bell's detailed and explicit write-up in the leading American physics journal, and Maiman's less forthcoming articles in British magazines, some initially assumed it was Bell Laboratories, and not Maiman, who had won the race.[83] Maiman's victory was only belatedly crowned with recognition, and even then there was wormwood among the laurel leaves (figure 3.10).

As it gradually became clear, over the course of the summer, that a pink ruby laser had been successfully operated, the laser teams

Figure 3.10
Theodore H. Maiman and Irnee J. D'Haenens in 1984 on the occasion of Maiman's induction into the Inventor's Hall of Fame. (Courtesy AIP Center for History of Physics)

already in existence intensified their efforts or changed their direction. Sorokin and Stevenson put aside their parallelepipeds. They had their CaF_2 boules made into cylinders with silvered ends, bought flash lamps, and in a period of sustained labor culminating over the 1960 Thanksgiving holidays, lased the uranium sample. Because it was a four-level laser, the pumping power required for lasing was orders of magnitude smaller than that required for the three-level ruby laser.

At Columbia, Townes moved Abella into a research program on ruby, while Cummins continued the cesium project. Snitzer and his coworkers, who had been using high-pressure mercury lamps, abandoned these and ordered flash lamps. They also briefly experimented with drawing ruby out into optical fibers and cladding them with glass.[84] Schawlow, who had joined the collaborative effort at Bell Laboratories to corroborate Maiman's results on pink ruby, went on to use the power supplies and flash lamps available from this work, together with the results of his recent spectroscopic studies on dark ruby, to exhibit lasing in dark ruby, in collaboration with his technician, George Devlin. Irwin Wieder had just moved from Westinghouse to Varian when Maiman's measurements of pink ruby's quantum efficiency appeared.[85] He was able to appreciate as well as anyone the importance of Maiman's result, but he was unable to get equipment for laser work until after the Hughes press conference. Once he did get his laboratory outfitted, he collaborated with Lynn Sarles, in a project independent of Schawlow's, to make red ruby lase. The Wieder-Sarles work appeared back-to-back with the Schawlow-Devlin paper in the *Physical Review Letters*.[86]

Javan, Bennett, and Herriott continued on the path they had laid out for themselves. By this time, Javan and Bennett were ready to try out the resonator Herriott had designed for the experiment. In the first trial, they attempted to pump a discharge tube with a powerful magnetron, but the quartz tube confining the gases melted. A new apparatus was built, but during the bake-out that was one of the steps needed to ensure a high vacuum, the coatings of the mirrors flaked and fell off. The experiment was reassembled and baked-out at lower temperature. This time the mirrors took on a linear pattern from the heat but did not fall off. On December 13, 1960, attempts were made to line up the mirrors and detect lasing. They were unsuccessful. Then, in the afternoon, Javan, who was watching Herriott idly rotating the micrometers that adjusted the mirrors, heard him suddenly say, "My God, what is that?" Herriott writes, "I was casually turning the micrometer on one of the mirror adjustments when a signal suddenly appeared on the oscilloscope .... We ... adjusted the monochrometer and found the signal peaked at [1.153 microns], one of the expected wavelengths. We then adjusted the mirror angles to maximize the signal. We found four other wavelengths that correlated with expected transitions. We then opened a bottle of 100 year old wine that Javan had bought in Paris."[87] The last of the 1960 lasers and the first using a gas medium and giving a continuous, rather than pulsed, beam had come into operation (figure 3.11).[88]

What had the group who made the 1960 lasers added to what Schawlow, Townes, and Gould had achieved by the end of 1958? First, they had made exhaustive studies of materials other than the potassium vapor Townes and Schawlow discussed, and they had

Figure 3.11
Donald Herriott, Ali Javan, and William R. Bennett jr. clowned for the camera alongside their helium-neon laser on December 13, 1960, after the official visitors left. Herriott holds a beaker of champagne. (Courtesy William R. Bennett jr.)

established the requirements for population inversion in these materials. Examples are Javan and Bennett's elegant theoretical and experimental study of the helium-neon system, Maiman's work on pink ruby, Schawlow's exploration of dark ruby, and Sorokin and Stevenson's investigation of the properties of U^{+3} and Sm^{+2} in a CaF_2 host. Second, they had found sources of pumping power that could meet the inversion requirements they had established. Maiman's solution, in particular, was both unexpected and influential. His bold decision to use photographic flash lamps was picked up immediately and made possible both the IBM calcium fluoride lasers and the dark ruby lasers. In contrast, and somewhat surprisingly, there was no advance beyond the Fabry-Perot resonant structure that Schawlow, Gould, and Townes had suggested. Despite the uncertainties a number of scientists expressed about it at the time, and despite the initial attempts of Sorokin and Stevenson to substitute prisms with total internal reflection and of Snitzer to work with clad glass fibers, all the operating 1960 lasers made use of Fabry-Perot structures. The proposals for Fabry-Perot resonators by Gould and Schawlow were vindicated both by the lasers' success and by the work of Fox and Li.

It is worth remarking on the fact that not just one but four different types of lasers had been made to operate. This variety is, of course, a manifestation of the fact that a number of different teams had entered the field and that each team had a slightly different strategy, one shaped by the experience and education of its leading members, their style of doing science, and the research environments within which they worked.

The influence of style is clearly visible in Townes's choice of potassium and Javan's choice of noble gases. The influence of experience is revealed in Maiman's choice of ruby, a material of which he had a thorough knowledge. Other things being equal, scientists can get a leg up by using a material with which they are already conversant.

The influence of research environments is in many cases unproblematic. That Snitzer should work on glass fibers in the context of the American Optical Company, or that Schawlow should focus on continuous-wave lasers as a member of the Bell Telephone Laboratories staff, seems a straightforward response to company needs.[89] Other examples are more tricky. Javan and Bennett worked in Bell's Research Area, but they also had academic orientations. Bennett had just come from a university, while Javan appears to have thought of Bell as a stepping stone to a university position.[90] Throughout two intensive years, Javan's and Bennett's research exhibited a double aspect—they strove to make a laser and also did elegant spectroscopy. It seems clear that such work was directed as much to the physics discipline as to AT&T. At Hughes, it was not unusual for scientists to have dual backgrounds, as Maiman did, in fundamental and applied work. This made them more sensitive to the practicality of new devices than were, for example, academics or members of industrial laboratories, such as Bell's research groups, that were more insulated from development. Maiman testified in 1967 that one reason he chose pink ruby was that he was attracted by its potential to operate at high powers and reasonable temperatures and to function in a rugged and compact device.

Townes at Columbia University worked on a potassium laser, for which predicted output powers were fractions of a milliwatt. Such a low-power laser would only be suitable for purely scientific work. To be sure, he initially chose potassium as the most secure route to a functioning laser, and he foresaw that other lasers, with other areas of application, would follow. In particular, Townes, given his close association with the Institute for Defense Analyses, had reason to be cognizant of possible military applications. But Townes's position as an academic scientist both reflected and motivated a strong commitment to the market of the discipline of physics, and the creation of a new scientific instrument was a reasonable product for this market.[91]

From all of this emerged a solid pulsed three-level laser, a number of pulsed four-level solid-state lasers, and a continuous gas laser. Collectively, they added up to a spectacular demonstration that the idea of the laser was sound.

4

Laser Research Takes Off

Introduction

Once launched, U.S. laser R&D made a nearly vertical takeoff. This explosion in effort from early 1961 though the end of 1963 is easy to demonstrate quantitatively, from a variety of statistical indicators. There were about 475 participants in the Second International Conference on Quantum Electronics, held in Berkeley in March 1961. There were 1,100 participants at the Third Paris Conference in February 1963.[1] *Physics Abstracts* listed (in rough numbers) 20 laser publications (out of about 21,000 abstracts in all) in 1961, 120 (out of 24,000) in 1962, and 270 (out of 26,000) in 1963. A bibliography of laser literature published in 1967 has 17 entries for 1960, 136 entries for 1961, 304 entries for 1962, and 752 entries for 1963.[2] Qualitatively, however, the laser boom was a complex phenomenon.

We must recognize to begin with that most scientists spend most of their time on routine work. The attraction, therefore, of a new field, where interesting and important phenomena lie around waiting to be found, is substantial. Many laser pioneers recall this attraction. "The sense of excitement was enormous," wrote one recently. "Anyone who worked on lasers in those days remembers it. . . . There is always a certain amount of pleasure when speculations and hypotheses are confirmed . . . but there is a different kind of excitement when something both new and significant is achieved."[3] Integrally bound to this *joie-de-recherche* was the chance to improve one's professional reputation. "[It was a way to get] a publication that somebody would read . . . sort of cream-skimming . . . . Clearly, whatever you did was going to be important. That's why it was so easy [for a manager] to get people to do things. You [didn't] have to twist their arm, just say, 'Here's a piece of the action.'"[4]

The enthusiasm that this undeveloped field kindled among scientists, however, is only a small part of the story. Another aspect is the reactions it provoked in the professional societies. One of the salient features of the laser is that it created fresh connections among old subjects. The maser had forged a new link between physics and electrical engineering; the laser took this a step further and tied in optics. This had the social concomitant that three scientific communities were now actively involved—physics, engineering, and optical science. The more that people who identified themselves as electrical engineers or optical scientists embraced laser research, the more of an effect it had on the professional societies to which they belonged. Laser conferences had to be organized, and laser articles had to be placed in journals. Which organizations were to sponsor the conferences? And into whose journals should the scientific papers go? The situation presented clear opportunities to the professional societies to recruit new members and invigorate their publications. It also presented opportunities for the discomfiture of members in traditional fields, for jurisdictional quarrels, and for tensions. Once a society decided to take lasers under its authority, it became itself a force for accelerating laser research. The society stimulated interest and promoted new ideas through meetings, solicited laser articles for its journals, and included lasers in its educational work in colleges and universities.

If lasers were "hot" in the early 1960s, they ascended to this temperature, above all, in the industrial sector of the U.S. research establishment. Analysis of the papers in *Physics Abstracts* shows this clearly. Industrial laboratories submitted 64% of U.S. laser papers listed in the 1961 volume, 78% of U.S. papers in the 1962 volume, and 71% of U.S. papers in the 1963 volume. These figures are the more impressive in that the listings in *Physics Abstracts* are biased toward academia and against industry. They do not include the vast "gray literature" of technical reports, unclassified and classified.[5] We can assume that this literature reports a higher proportion of the work done in industry than in universities, since industrial scientists are under much less pressure to publish in the open literature.

Managers at industrial laboratories had a variety of reasons for steering their technical staff toward lasers. Some foresaw that lasers might eventually underlie the next generation of their company's technology. This was the case at Bell Laboratories, where managers saw optical communications as the next possible step after satellite communications. Some deemed lasers a better way to do a job at hand. Scientists at Martin-Marietta's Orlando Aerospace Division, for example, had been developing a target designator that used incoherent ultraviolet light; once the glass laser had been made to work, they switched their focus to a laser target designator, which they expected to be more compact and to have a longer range. Other companies, such as Hughes and Westinghouse, were motivated by the evident interest of the Department of Defense; they saw the possibility of substantial sums for R&D contracts and large procurement contracts farther down the road. And some companies simply wanted to keep a hand in, since lasers appeared to be an important new technology and research was, in the early 1960s, an "in" activity. *Aviation Week and Space Technology* estimated in 1962 that about 400 companies had some sort of laser research program in progress and that 10–12 of these were substantial.[6]

From the start, industrial laboratories mounted projects on laser systems as well as projects on devices. Already in late 1960, Hughes Aircraft Company had two independent teams building optical radars; both had "breadboard" models operating by the end of January 1961. American Optical, whose interest in the medical market grew out of its historical involvement in ophthalmic lenses, had a research program on laser photocoagulators for retinal surgery by 1961. Between systems and device research a synergism developed, for the attempt to design or build systems with a component as underdeveloped as the laser inevitably led scientists to pose new, often well-defined, demands for improvements of the laser itself. Moreover, the richness of previously unknown effects produced within this new instrument, and by the beam it emitted, was so great that the very activity of exploring applications, by creating a situation in which additional people worked with lasers, ensured the discovery of new phenomena that would require fundamental investigation and explanation.

Business did not, however, limit itself to research on systems, devices, or laser phenomena. As early as 1961, companies began placing lasers on the commercial market. Some of these were large firms such as Hughes or Raytheon that had the resources to field substantial research programs. Some, such as Trion Instruments

Inc., were garage businesses with a handful of part-time owner-workers. And some were new companies dedicated to lasers from the start by the scientists who founded them; in this chapter, we will examine Korad as a scientist-entrepreneur firm of this type.

The dollar value of commercial sales was tiny. *Barron's* magazine estimated annual sales of $1 million in early 1963. Sales alone, however, do not measure the stimulus the laser industry gave the field. The pace of research in industry, government, and universities alike was quickened by the availability of off-the-shelf lasers. Even in the earliest years, some seminal experiments used lasers from commercial houses. A 1961 experiment at the University of Michigan that opened the new field of nonlinear optics used a Trion laser.[7] The "Luna See" Project at MIT, which in 1962 sent light to the moon and detected the echo, used a Raytheon laser.[8]

Laser manufacturers contributed their share of improvements in laser tubes, mirrors, and flash lamps as they struggled to bring the wayward early instruments up to marketable standards. They also contributed ideas for new applications that might have the potential to expand their markets. Industry programs, both those devoted to investigating lasers in the laboratory and those producing lasers for sale, also had a directly stimulative effect on university interest because the companies hired academic consultants. It is likely that most academic scientists working on lasers in the 1960s held consultancies.[9] At times, consultancies simply enabled university scientists to do more of the laser research they had already embarked on, but in at least one case, as we shall see, a consultancy stimulated a major and extremely fruitful line of university research.[10]

Military interest was the warp through which the woof of other U.S. effort was woven. Each service had its own slant. The Air Force (and especially the Aeronautical Systems Command and its laboratories at Wright-Patterson Air Force Base, which soon became lead laser groups within the service) was interested in sun-powered lasers, laser space communications systems, and systems for the identification and tracking of space objects. The Army foresaw a use for laser beams as guides along which missiles could home in on their targets ("target designator" systems). It also wanted high-energy beam weapons for both antipersonnel and antimissile missions. In Army weapons work, the Army Ordnance Missile Command and its laboratories in Huntsville, Alabama, took on the major role. As noted in chapter 3, the appeal of laser beam weapons was one factor that convinced ARPA to fund TRG's research.[11]

The rapid progress that laser technology made during 1961 heightened the military's excitement over lasers. One important development was Elias Snitzer's success in creating a glass laser at American Optical Company. As we saw, Snitzer had originally been attracted to glass fibers because fibers can readily sustain electromagnetic modes. His research group, which had initiated intensive work in the summer of 1960, had tested about 200 fibers with different compositions and had even made a fruitless attempt to draw ruby into a fiber and provide it with a cladding. These were all fibers doped with substances known to fluoresce at visible wavelengths. By the end of 1960, though, both of Snitzer's coworkers had transferred to the classified Air Force–sponsored project to create a sun-powered laser transmitter. Snitzer continued on alone. At this point, he decided to shift from visible to infrared fluorescers. This meant using a different group of dopants: The infrared-emitting glasses would be those containing the rare earth elements neodymium, praseodymium, holmium, erbium, and thulium.

By this time, Snitzer had also weaned himself away from fibers. The qualities he sought in a specimen were good transparency to the pump light and total internal reflection of the light that hit the sides as it traveled down the specimen. Transparency depended not merely upon physical thinness but also upon the absorption properties of the material. In the infrared fluorescers he was now considering, the absorption was low enough that he could switch from fibers to thin glass rods. Total internal reflection, he had also come to appreciate, was a matter not of size but of choosing an appropriate cladding material for the rod and securing a close conjunction between the core rod and the cladding. This meant that glass lasers could theoretically be of any dimensions, and because glass boules could at that time be created in much larger bulk than ruby crystals, glass lasers promised to deliver much higher total energy. Glass could also be made with better optical quality than ruby, and that promised less spreading of the beam. In October 1961, Snitzer succeeded in lasing a barium crown glass doped with trivalent atoms of neodymium.[12]

In December 1961, William Culver, a scientist at the Institute for Defense Analyses, reported to ARPA that the improvement of ruby lasers and the invention of the glass laser had led a number of people in industry and government to the view that a practical beam weapon might indeed be possible. ARPA convened a special panel of prominent scientists, which met at the end of 1961 and recommended a high-priority research and development program. The program, called Project Seaside, was administered by the Office of Naval Research. It concentrated on ruby and glass lasers and associated technologies.[13] Its inauguration in 1962 shows up as a substantial spurt in the level of DoD funding for laser research.

A rough estimate for DoD spending for extramural laser research in 1960, based on figures in the trade publication *Aviation Week and Space Technology*, is perhaps $1.5 million. Of this amount, $750,000 was for a renewal contract to TRG, and $70,000 for an Air Force contract awarded to Hughes Research Laboratories in the wake of Maiman's ruby laser. In 1961, the department spent about $4 million.[14] In 1962, DoD laser spending reached $12 million, about half of all laser R&D money disbursed during the year in the United States. For fiscal year 1963, *Aviation Week and Space Technology* estimated that DoD expenditures, including intramural projects, were between $19 and $24 million.[15]

Through ARPA and direct service contracts, military funding supported most of the university projects. Of the 51 laser papers listed in *Physics Abstracts* for 1961–1963 that came from universities, 41 or 80% acknowledge at least partial DoD support.[16] Industry also did much of its laser research and development on military money. *Aviation Week and Space Technology* estimated in March 1962 that the U.S. military was funding more than 75 industrial laser R&D projects. In April 1963, it counted almost 130 such contracts.[17] Private firms that committed their own resources often intended their input as seeding in the hope of eventual military R&D or procurement contracts. The very existence of small laser firms often depended upon the fact that Department of Defense R&D contracts were easy to obtain. Money from DoD contracts provided a vital source of funding even for those fledgling companies, such as Spectra-Physics, that intended eventually to establish themselves in the civilian market. Some of the most stimulating of the laser conferences sponsored by academic institutions and professional societies were actually initiated, as well as funded, by military agencies.

The laser beam weapons program was seized on by the popular magazines. The laser had already struck them as newsworthy, and the idea of a weapon firing light, "a kind of warfare so far envisaged only in science fiction," as a *Reader's Digest* article pointed out, gave the topic a nice twist.[18] The military's enthusiasm was picked up by Wall Street. Financial journalists predicted in 1963 that the laser industry, buoyed on a wave of military spending, could reach an annual sales of $1 billion within ten years.[19] This optimism in the financial press translated into a greater availability of venture capital and thus directly aided the small commercial firms.

The laser, then, held something for each of many sectors of the U.S. technical community. For the scientist and engineer, it was a source of work that was both intellectually engaging and useful for advancing careers. For professional societies, it provided a vehicle both for rendering service to their disciplines and for building their membership and organization. The laser promised the military new tactical and strategic weaponry, and it held out the prospect of civilian applications such as medical therapies, computer components, and more capacious communications links. It gave investors growth securities and scientist-entrepreneurs the chance to get in on the ground floor of an industry that was expected to develop rapidly. The press had another scientific marvel to tout in an era in which interest in science, and especially in scientific achievements that contributed to national prestige, was high. Each of these sectors egged on others. The military supported academic work. Press articles titillated investors. The search at large firms for laser applications provided a market for the commercial lasers of small firms. And so on. A circle of interacting interests was set into motion, a pinwheel of excitement in the landscape of the early 1960s.

Lasers and the Professional Societies

In 1960, the professional societies in the United States associated with atomic and molecular physics, optics, and electronic engineering were, respectively, the American Physical Society, with a membership of about 16,000, the Optical Society of America, with a membership just under 3,000, and the Institute of Radio Engineers, with a membership of 72,000.[20] Of these three groups, the Optical Society perhaps stood to gain most from the glamor surrounding the laser, for while optical science appeared to its practitioners to be undergoing a renaissance, to the rest of the physics community it looked like a backwater.

The field served by the Optical Society was a broad one, embracing among other things spectroscopy, photography, lens design, the study of color, the physiology of human vision, and a rich variety of optical instruments and optical computer devices. The military buildup of the 1950s had stimulated optics as it had electronics. The Department of Defense had a long list of optics needs including surveillance cameras, "night vision" instruments for carrying out air reconnaissance and other operations in conditions of very low light intensity, infrared sensors, missile trackers, and periscopes. In World War II and after, microwave radar had cut into optics' traditional military uses to some extent, because microwaves, in contrast to light, could be propagated through the atmosphere under all weather conditions. But the massive post-Sputnik move into space was opening up a whole new field of extra-atmospheric applications for optical instruments. Stimulus had also come from such new technologies as fiber optics, thin-film optics, and the integration of optics and electronics into "optoelectronics."

One result of the field's growth was that the Optical Society was expanding. Its 1960 membership was nearly five times its 1940 membership. It had opened a permanent office in Washington, D.C., just the year before and had appointed its first full-time executive secretary, Professor Mary Warga, a spectroscopist from the University of Pittsburgh. The society had also approved a new journal, *Applied Optics*, to supplement the venerable *Journal of the Optical Society of America* (*JOSA*). *Applied Optics*, begun in 1962, was to be more topical and lively than *JOSA*, which would now be reserved for research papers. It was to offer an outlet for the articles on optical engineering then appearing in a wide variety of magazines.[21]

Despite the field's growth, optics as it was practiced in 1960 did not make much contact with the intellectual frontiers of physics. Optical scientists did not deal with quantum mechanics or quantum field theories, nuclear physics, or high-energy physics, all central concerns of physicists since the war. Hence in the constant process of negotiation among faculty members over curriculum that is a feature of American physics departments, optics courses were being squeezed out to make room for more stylish subjects. While 7.8% of employed physicists were working in optics in the early 1960s, only 1.5% of university graduate students were specializing in optics.[22]

The skills of the optical scientists were vital to the laser enterprise. We have already seen the role optics expert Donald Herriott played in the Bell Laboratories helium-neon laser program. Townes, when he took his leave from Columbia to serve as vice president of the Institute for Defense Analyses, chose a British expert in optical thin films, Oliver S. Heavens, as his stand-in for the spring 1960 term. The mirrors and windows of lasers and the systems of lenses to guide and deliver the beams all depended on optical engineering.

At the same time, the laser in 1959 and 1960 presented scientific puzzles difficult enough to command the respect of the most sophisticated physicists. One of these puzzles was coherence. The whole point of the laser was that it would produce coherent light, similar to the coherent radiation radio and microwave sources produced at lower frequencies and differing from the random radiation produced by incandescent bulbs or neon lights. But what is coherent light? There was no consensus on that point. In fact, the whole field of coherence had been thrown into an uproar a few years earlier by the work of a pair of British scientists, Robert Hanbury Brown and Richard Q. Twiss. The two had invented a technique, called "intensity interferometry," for measuring the angular diameter of radio stars, and in 1954 they proposed applying it to optical astronomy.

The proposed optical intensity interferometer would use two detectors separated by up to 200 meters and focused on a single star. The light received at each detector would be converted into an electrical current, the two currents would be brought together, and from measurements of their "correlation" scientists would calculate the stellar diameter. The fact that a correlation existed between the currents implied that the spots of starlight hitting the two detectors were coherent. There had been no objections to the technique when it was applied to radio stars in the early 1950s, but there was a storm of criticism when the experiment was proposed for visual stars. The problem was that electromagnetic radiation must be describable in terms of photons as well as waves, and in the visible region of the spectrum the particle properties (photon features) become more significant. The kind of correlation between photocurrents that Hanbury Brown and Twiss were assuming struck many physicists as a violation of the properties quantum mechanics predicted for photons. Their proposal therefore opened a can of puzzles about how coherent photons actually behave and the way in which a photodetector preserves in its electrical output accurate information about incoming photons.[23]

The topic of coherence was clearly a good meeting point for the small contingent of the optical community interested in the application to optical instruments of classical theories of coherence (that is, theories based on the wave properties of light) and the physicists who were struggling to make working lasers. Yet it was not simply a matter of optical scientists and laser scientists gravitating together. As so often in our story, military organizations and industrial companies played roles as matchmakers between the scientists and the new research topics.

One example of this matchmaking was the Conference on Coherence Properties of Electromagnetic Radiation, held at the University of Rochester in June 1960. This was an important forum for bringing the laser to the attention of optical scientists. It was initiated by the Physics Division of the Air Force Office of Scientific Research. Some of the thinking behind the initiative is indicated in a memorandum circulated in April 1959 by William S. Rodney of the Physics Division. He called attention to the problems with the concept of coherence that had arisen in the context of "the present effort to produce radiation at optical frequencies with coherency similar to that of microwaves" and were connected with the "difficult concept" of coherence or correlation between photons emitted from different atoms. Meanwhile, the concept of coherence in optics had only been used to discuss the propagation and interference of light beams, while "the radiation from two different atoms is considered to be completely incoherent." Coherence in lasers and coherence in classical optics had so far failed to make contact. Rodney suggested that "we try to stimulate more basic research in the field of coherency and photon optics under the general title of 'Photonics.'"[24]

Rodney had already approached Robert E. Hopkins, director of the University of Rochester's Institute of Optics, with the idea of organizing a conference on coherence. The Institute of Optics was at the time the preeminent U.S. center for the training of optical specialists. It was one of a number of optical institutes set up following World War I, when it had become clear that optical competence was vital to outcomes on the modern battlefield.[25] Hopkins, himself a specialist in lens design, was taking steps to strengthen the institute's capabilities in basic optical science. As one avenue toward that goal, he was negotiating with the Czech-born British physicist Emil Wolf, who was then just finishing his collaboration with Max Born on *Principles of Optics,* a textbook that serves to this day as the authoritative compendium on classical optics.[26] Wolf was an expert in the classical theory of coherence, and he was already wrestling with the coherence questions implicit in the experiments of Hanbury Brown and Twiss; these were research topics that Hopkins viewed as important. Rodney's conference proposal made the idea of hiring Wolf even more attractive. Hopkins welcomed it for other reasons as well. It would bring in some money, even if only for a single conference; it would enhance the institute's image; and it would connect the institute with the kind of advanced topic Hopkins sought to foster. It was agreed that the institute would host the conference and that Wolf, who arrived at the university in April 1959, would chair its program committee. In February 1960, Rodney went to the Optical Society of America and secured its assent to join the Air Force and the university as the third cosponsor.[27]

The conference amply met the objectives of both its Air Force and university sponsors. It assembled a distinguished international group of university, government, and industry scientists. The participants represented just the kind of specialties William Rodney had sought to bring into contact: spectroscopy, quantum electronics, radioastronomy, optical theory, and fundamental physics. Hanbury Brown discussed the experiments he had carried out with Twiss, and electrical engineer A. Theodore Forrester presented results of an experiment made prior to the Hanbury Brown–Twiss work on light but largely ignored until that controversy indicated its importance.[28] Townes was represented by Oliver Heavens. Dicke, Weber, and Snitzer all spoke. The classical coherence theory and its connection with optical instruments was well covered by Wolf himself, by his friend Leonard Mandel from Imperial College of the University of London, and by researchers from institutions as distant as Bangalore, Sydney, and Paris and as near as the Hudson Valley.[29] The University of Rochester thus became inextricably connected with the fundamental and provocative concept of coherence (the conference in fact became known as the Rochester Conference on Coherence) at the best possible moment: The meeting took place from June 27 through June 29, 1960—less than two weeks before Maiman's electrifying announcement brought the idea of coherent light so prominently before the scientific and lay public.

Nor was William Rodney disappointed in his hope of stimulating "more basic research in the field of coherency." One optical scientist who turned his attention to laser problems was Wolf himself. Prior to coming to Rochester, Wolf had focused his research on partially coherent light, which was the type of light then available in laboratories for use with optical instruments. But the laser's degree of coherence was astonishing. Using a quantitative measure of coherence he called the "degeneracy parameter," Mandel estimated that the best previous laboratory sources of light had degeneracy parameters of about 10^{-3}, whereas the ruby laser had a degeneracy of 10^7 and the Javan-Bennett-Herriott helium-neon laser, just coming into operation, had a degeneracy parameter of 10^{12}.[30] Wolf, with contract money from the Air Force Office of Scientific Research, obtained through his contact with Rodney, began devoting some of his time to extending classical coherence theory to the case of highly coherent light.[31]

The papers Wolf began to write—many in collaboration with Leonard Mandel—illustrate some of the thinking in the early 1960s. To this point, many physicists and engineers outside the specialty had only a limited acquaintance with optical coherence theory. Mandel and Wolf were obliged to point out that coherence was not equivalent to monochromaticity;[32] therefore, equations that had been applied to laser beams simply because they were valid for monochromatic light had to be examined anew. Conversely, laser light could serve as a touchstone, to show which "established" results within the old coherence theory had to be reinterpreted or extended.[33] Wolf's early papers also demonstrate that the basic question of why a laser light beam is coherent was still unanswered. For this reason, Wolf in 1963 posed the question of whether spatial coherence resulted from the interaction of the light with the atoms in the cavity, or whether it comes about instead because of the repeated reflections the light undergoes at the mirrors. Nor was there much clarity on the extent to which laser light's properties resembled those of ordinary thermal light.[34] Thus a subject as fundamental to physics as the nature of light was revealing itself as ill-understood.

Wolf came to the study of the coherence properties of lasers from the side of optical science. Another major figure who also came from outside the community of quantum electronics was Roy J. Glauber, a professor of theoretical physics at Harvard University who had worked in quantum field theory and nuclear physics. Glauber began his coherence studies at the request of Saul Bergmann, a laser physicist from the Research Department at American Optical Company. Bergmann wanted to understand the relationship between the work of Hanbury Brown and Twiss and the properties of laser light, and he invited Glauber to help in this project as a consultant to the company. Glauber had not himself been doing research on optical coherence, but he had been in touch with it because some of the theoretical and experimental work was being done in the Harvard physics department.[35] He now took as his point of departure the question: How can Hanbury Brown and Twiss's results, and the completely coherent character of laser light, be embodied in a fully quantum-mechanical theory? The solution, as Glauber began to perceive it in 1962 and early 1963, lay in extending the notion of coherence, which until then had been expressed mathematically through the use of quadratic functions, to include quartic functions and functions of higher power ("higher-order correlations"). The highly monochromatic thermal light that Hanbury Brown and Twiss had used in laboratory experiments to test the principle of their proposed optical intensity interferometer had had only first-order (quadratic) coherence. Laser light had a much higher order of coherence, and by examining the properties of higher-order correlation functions, Glauber concluded that laser light would show no intensity correlations of the sort that Hanbury Brown and Twiss proposed to use as the basis of their stellar diameter measurements.[36]

February 1963 is a date we could choose for the beginning of the integration of these coherence studies into laser science. In that month, Wolf, Mandel, and Glauber were among a number of speakers on coherence topics at the Third International Conference on Quantum Electronics in Paris.[37] Yet we would have to emphasize the word *beginning*. The theories were still in evolution. A harsh disagreement over the relative merits of classical versus quantum-mechanical approaches had begun to take shape between Wolf and Mandel (soon joined by E. C. G. Sudarshan) and Glauber.[38] And coherence theory, even when it became elucidated in the course of the 1960s, held low priority for many laser workers. It was not needed to make lasers work, and it was hard to master. This was perhaps particularly true of Glauber's theory, which was full of beautiful insights into the structure of physics but, despite the clarity of Glauber's prose, was generally found to be exceptionally difficult. Well into the 1960s, some laser physicists had not learned the lessons of coherence theory and were still, for example, identifying "coherence" with "monochromaticity."

The Optical Society of America, meanwhile, was moving to provide a forum for the new field. It rearranged the program of its semiannual meeting in October 1960 to schedule a major talk by Theodore Maiman. That was the only presentation on lasers at the meeting. At the society's next meeting, however, held March 2–4, 1961, in Pittsburgh, OSA placed four optical maser symposia on its program, with four long papers and fourteen short ones. In addition, Townes gave one of the keynote addresses, a talk on "Optical and Infrared Masers." Most of the major players were represented with papers: Columbia University, Bell Telephone Laboratories, IBM, TRG, Hughes, Westinghouse Research Laboratories, and Varian (that is, Wieder and Sarles). *JOSA* reported that "interest in the papers on optical masers was particularly high. [Attendance at] the first invited paper . . . by Charles H. Townes . . . filled the ballroom of the Penn-Sheraton Hotel and overflowed into the balcony. . . . Similar attendance problems [at the other laser sessions] were handled by providing a separate room where those unable to gain entrance to the meeting room could hear the papers through a loudspeaker system."[39] One of the attendees, Peter A. Franken, then an associate professor of physics at the University of Michigan, gave a more colorful recollection: "Normally if you had a paper at the Optical Society, you might draw a hundred people. There might be two or three . . . cameras taking pictures of the slides. These halls were packed, the ballroom was packed, for these papers. I remember as a high point Art Schawlow getting up to give a talk. Every slide he projected, there was a veritable staccato machine-gun fire of Minoltas going off. It was unbelievable! Panicsville."[40] The interest in lasers shown at this meeting was echoed a few weeks later in Berkeley, when a follow-up to the September 1959 Quantum Electronics Conference was held. Like the first conference, the meeting at Berkeley was restricted to people who were actively doing research in the field. It drew 475 participants, as opposed to the roughly 180 who had attended the first meeting.

The Optical Society's new journal, *Applied Optics*, was a natural outlet for laser papers, and editor John Howard made them the central feature of the inaugural issue in January 1962. The number of papers submitted so far exceeded expectations that the journal issued a supplement on "Optical Masers" at the end of 1962. Subscriptions also ran above anticipations. More than 3,300 were received by March 1962, at a time when the society had a membership only about that figure. The special supplement of 4,000 copies sold rapidly. Within its first year, *Applied Optics* was advanced by the OSA board of directors from a bimonthly to a monthly. While the magazine had appeared at an opportune moment for laser scientists, the reverse was also true: *Applied Optics* was riding the crest of the tumult the laser was creating within the scientific and engineering community.[41]

Masers had, of course, been outside the purview of the Optical Society. Nor had any of the society's members participated in the initial work of formulating the idea of the laser or in constructing the earliest operating models. After 1961, however, the society began to bring laser practitioners into its membership and into its governing structure. Arthur Schawlow, one of the first laser researchers to turn up on the society's committee lists, remembers that Mary Warga had been convinced that the field of lasers should be given an important place in the society's activities. "She came up to Bell Labs and had lunch with a number of us, and persuaded us to join the Optical Society, and then made sure that a number of us were invited to give talks at meetings. . . . [It was] Mary Warga particularly [who] got me to join."[42] At the Institute of Optics of the University of Rochester, Robert E. Hopkins also moved decisively to bring in laser research, appointing three experimentalists to his staff.[43]

U.S. optical organizations and laser science had begun to enter into a relationship of mutual aid. The Optical Society was becoming a major forum for the presentation of results, and the Institute of Optics an additional site for research. Lasers, for their part, were contributing to the growth of the society and the availability of jobs for its members. They were associating the society with a field that excited scientists for the basic puzzles it posed as well as for its possibilities for application. It was, moreover, a field that the military was promoting. This meant that military funds were available for direct support, for conferences and such, and also for indirect support, since the money the military was pouring into university and industry research contracts translated into an increased volume of results to be reported at meetings or in the pages of society publications.

The much larger Institute of Radio Engineers (IRE) had a far more articulated organization than the Optical Society. A system of semiautonomous professional groups, each built around a technical area, had been instituted after World War II to accommodate the succession of new electronic specialities that was emerging. The IRE was also on the verge of concluding a merger with the older but smaller American Institute of Electrical Engineers.[44] From October 1961, when the merger was voted in, until the two societies joined in the new Institute of Electrical and Electronic Engineers (IEEE) in January 1963, an even more searching examination than usual was devoted to the scope of the professional groups and of the *Transactions* journals that were their organs of publication. Within the IRE/IEEE, therefore, the assimilation of the laser became a matter of determining the professional group that should have jurisdiction over it and the appropriate journal within the organization for laser publications.

Both the Professional Group on Electron Devices (PGED) and the Professional Group on Microwave Theory and Techniques (PGMTT) laid claim to lasers.[45] PGED had under its jurisdiction tubes such as klystrons, magnetrons, and traveling-wave tubes that generated electromagnetic radiation, and the PGED Administrative Committee held that lasers, which also generated electromagnetic waves, albeit of a different frequency and by a different physical effect, were an extension of such tubes.[46] PGMTT Administrative Committee members pointed out that masers were clearly microwave devices and that lasers were an extension of masers. The PGMTT asserted rights to the laser with less unanimity, however, since it was also concerned that an overemphasis on lasers might compromise the group's ability to serve its other specialities. Meanwhile, outside the PGED and PGMTT, other IRE members were petitioning the Professional Groups Committee to organize a fresh professional group dedicated to quantum electronics.[47] This was a move that both the PGED and the PGMTT opposed.[48]

One argument that proponents of a Professional Group on Quantum Electronics made was that no journal within the organization yet served as an effective vehicle for laser articles. Both the PGED and the PGMTT had been making continuing efforts to get laser engineers to publish in their respective *Transactions*, but their efforts had met little success. Instead, laser engineers were publishing in the letters section of the *IRE Proceedings*, in the American Physical Society's *Journal of Applied Physics* and *Applied Physics Letters*, and in *Applied Optics*. In 1964, the PGED and the PGMTT each made plans to increase the number of laser papers in its own *Transactions*. The PGMTT appointed W. Kahn as an associate *Transactions* editor for lasers, and the PGED appointed Eugene I. Gordon to a like position.[49] PGED *Transactions* editor Glen Wade and associate editor Gordon then laid plans for a special laser issue for their magazine. At this point, late 1964, F. Karl Willenbrock of the IEEE Publications Board intervened. A commercial publisher, Pergamon Press, had been sounding the market with the idea of a quantum electronics journal. Willenbrock, determined that the IEEE not be scooped in this manner, negotiated to elevate the PGED *Transactions* special issue into the first number of a new PGED-sponsored *Journal of Quantum Electronics* to begin publication in the spring of 1965.[50] Jurisdictional problems were resolved in the middle of 1965 by the reestablishment of the journal under a joint PGMTT-PGED Quantum Electronics Council, with Gordon as its first chairman.[51]

The tight meshing of concern for laser science with more narrow and parochial concerns for building the IEEE, or some subgroup within it, is illustrated in Gordon's recollections of his role in founding the *Journal of Quantum Electronics*: "Electrical engineers were not interested [in lasers], and those who were interested were not publishing in the electrical engineering journals.... Obviously, there's a sort of chauvinism.... I [had] cast my lot with electrical engineering, not with physics.... So since I [had gotten] into the laser business, I was going to get electrical engineers in IEEE into the laser business, and then to make it successful, we obviously had to attract physicists.... This was really an electrical engineering enterprise ... [but] you've got to have the expertise, and the only way that expertise is going to get into electrical engineering in a reasonably fast way was through the physicists. So my goal was to get physicists to publish in the *Journal of Quantum Electronics* and thereby get them into IEEE."[52]

The Commercial Laser Industry

Twenty to thirty U.S. firms had lasers on the market by the end of 1963. Entrants to the new industry included both small scientist-entrepreneur companies and more established, multiproduct businesses.[53] A writer for *Barron's* magazine estimated at the start of 1963 that laser sales were running at about $1 million a year, with a total of a few hundred units sold so far.[54] This is vanishingly small compared to the total sales of electronics products, which were close to $14 billion in 1962.[55] Moreover, the figure of 20–30 commercial companies is small compared with the 400–500 U.S. companies that had entered laser research. Nevertheless, commercial laser production is interesting because the way in which the emerging laser industry stimulated other sectors of the technical establishment tells us something about the interactions that intensified the laser boom.

The outlook for new electronics ventures was excellent in the early 1960s. Venture capital was available from big companies with extra cash on hand, from investment bankers, and from individual investors.[56] The industrial electronics market had been growing rapidly and promised to continue expanding indefinitely. The military and defense budgets were also growing. The Kennedy administration, which took office in the spring of 1961, was committed to an acceleration of space research and a buildup of the U.S. military establishment, and the electronics industry could look forward to a healthy fraction of the funding increases.[57] Prospects got even better in late May 1961, after the Soviet Union had put the first human into space and recovered him successfully. Kennedy revised his space plans, moving from a mere substantial budget increase to a crash program to place men on the moon. *Electronics* magazine asserted that "a vast new astroelectronics market has been firmed up."[58] Electronics firms were also encouraged to field lasers by the large amount of R&D money available to university, industry, and government laboratories. This meant that laboratory scientists had the wherewithal to purchase a glamorous new device even in the absence of any clear plan for using it.[59]

Small electronics businesses enjoyed a special advantage at this time. The Kennedy administration believed in using government procurement practices, as well as fiscal policy, to reinvigorate the economy and was committed to giving small businesses a bigger piece of the defense pie. This reinforced the tendencies already resulting from the fact that the Department of Defense was procuring small lots of sophisticated electronic equipment. As a Battelle Memorial Institute publication explained, "This encouraged many small companies to be formed to serve extremely specialized defense markets. Scientific and engineering brainpower and some entrepreneurial desire were all that was needed to launch a business. Conventional mass-production, marketing capabilities and adequate capital were not needed. A high percentage of R&D funding by government was necessary and forthcoming."[60]

I shall describe here three small firms founded in 1961 and 1962, each with a somewhat different background. Trion Instruments, Inc., was founded in March 1961 by student-technicians at the University of Michigan's Willow Run Laboratories specifically to manufacture lasers. Spectra-Physics was started in September 1961 by five people who simply wanted a company of their own. It fixed on lasers some months later through the intervention of an optical firm that desired a laser arm. Korad, founded in November 1962, represents the classic case of inventors who leave the large laboratories where they did pioneering work in order to build businesses of their own around their inventions.

Trion Instruments was the brainchild of an inventive and unconventional Michigan student, Lloyd G. Cross, who had mastered the art of making ruby microwave masers while working as a technician in Chihiro Kikuchi's group at Willow Run. In the late 1950s, Kikuchi had recommended Cross for a contract to build a ruby maser for the Applied Physics Laboratory at Johns Hopkins University. Cross had organized a little company with some of his former classmates from Flint Junior College, so as to effect a legal separation of his commercial work for Johns Hopkins from his work for Willow Run. There was not much market for ruby masers, though, and soon after Maiman's discovery, Lloyd Cross turned to the manufacture of ruby lasers. With some other members of his original company, he founded Trion in March 1961. The founders capitalized the firm with $30,000 from friends and relatives. Some time later they succeeded in selling $100,000 in stock to an independent investor.[61]

Trion's sales of lasers averaged a few units a month. Nevertheless, it had entered the field so opportunely that in 1961 and 1962 it was one of the major U.S. producers of commercial lasers. Trion's list of buyers and renters for 1961 includes most of the major industrial and university laser programs. IBM, Texas Instruments, Bell Telephone Laboratories, Philco, Westinghouse, and Raytheon were all among the industrial customers; the University of California at

Berkeley, MIT, the University of Chicago, Ohio State University, and the Medical College of Virginia were among the academic customers.[62]

The instrument Trion's customers received was far from trouble-free. Elsa Meints Garmire, then a graduate student at MIT under Townes, recalled that she had to spend a year, in close interaction with the company, redesigning her laser. Michael Bass, who was working on his PhD at the University of Michigan under Peter A. Franken, remembered, "you'd go about 30 or 40 pulses and the coatings would have fallen off, and the system would stop lasing, or you'd run out of nitrogen and the alignment would shift."[63] Problems like these, however, were not unique to Trion but were typical of commercial lasers in a national environment in which commercialization of a new electronics device could begin immediately after its invention and proceed in parallel with hardware development and the growth of scientific understanding. That these lasers nevertheless facilitated research is illustrated by the case of Peter Franken's work on nonlinear optics.

It was the March 1961 meeting of the Optical Society of America that stimulated Franken to start his research. He recently recalled: "As I sat listening to [Schawlow's talk], I kept hearing, 'Lasers are fantastic—eye surgery, communication, eye surgery, communication'—eye surgery and communication were all anyone could think about. . . . So I began to think . . . there's got to be something unique here, and I calculated the strength of the electric field in an optical laser beam . . . and I discovered, holy cow, these are electric fields like 100,000 volts per centimeter! . . . Atomic electric fields are on the order of 10^9 volts per centimeter [so this] was getting to within 10^{-4}, 10^{-3} of an atomic electric field. That's got to be able to do something important for atoms, because I already knew that fields much less than that could auto-ionize atoms [the Auger effect]. . . . I did a crude estimate of [what would happen] if you took a laser and you focused it into any material . . . I was thinking then I'd use quartz, and I knew what the atomic energy levels were and I could make the calculations that there should be enough radiation produced to be detectable at the second harmonic."[64]

When the electric field of a beam of light traverses a transparent substance such as quartz, it excites an electric vibration in the quartz, and this in turn becomes the source of a secondary light beam. The frequencies of the secondary beam correspond to the frequencies of the quartz vibration. In ordinary light beams, the electric field is weak and the frequency of the excited vibration matches the incident light's frequency. Stronger incident fields, however, can incite a vibration that is a combination of the incident frequency with its overtones, or harmonics. Franken's estimate showed that laser light was so intense that a harmonic representing a doubled frequency would show up in the secondary beam. It should therefore be possible to use laser light at one frequency to create a beam of light at twice the frequency.[65]

The general idea of using laser light at one frequency to create light at other frequencies was not itself new. Even when lasers were only gleams in a few scientists' eyes, it was foreseen that two lasers of close but distinct frequencies could be used to produce a third beam whose frequency would be the difference between those of the initial two. In the 1940s, A. T. Forrester had suggested creating electromagnetic waves in the region from 1 millimeter down to about 0.05 millimeter by mixing light of different frequencies, and in the 1950s he and his collaborators had succeeded in demonstrating this combining process with incoherent light. Directly after the first lasers were announced, a number of concrete proposals were put forward for doing the same thing with lasers.[66]

Nor was the idea new of using a single electromagnetic beam and doubling (tripling, etc.) its frequency to give the second (third, etc.) harmonic. Microwaves, for example, had been doubled as another route, this time from the long wavelength end, to the submillimeter region. What was novel, however, was the mechanism that Franken was proposing for carrying out the doubling. Previous proposals for mixing laser beams had called for nonlinear devices such as photoelectric cells. Franken, in contrast, realized that even usually "linear" crystals such as quartz could be rendered nonlinear by the extraordinary intensity of laser light. He was excited enough to leave the meetings abruptly and return to his laboratory. There he refined his ideas in discussion with Gabriel Weinreich, a solid-state physicist, and enlisted the help of C. Wilbur Peters, an experimental spectroscopist, and Alan Hill, an undergraduate student who was a skillful experimenter.[67]

Franken and his team rented a laser head from Trion.[68] The instrument was marginal in terms of its intensity; it had a peak power of 3 kilowatts, in contrast to the 10 kilowatts Franken had counted on in his calculations. But it saved the several months of labor it would have taken to construct a laser. The Michigan team succeeded in demonstrating a small but detectable second-harmonic beam by the middle of July 1961. Despite the fact that, in one of physics' comic moments, the spot blackened by the second-harmonic beam was airbrushed out of the photograph of their spectrographic plate by a zealous lithographer,[69] their paper, published in the *Physical Review Letters*, had a pronounced and immediate impact. At one and the same time it opened a new field of scientific inquiry (the nonlinear properties of optical media) and a new field of practical technology (utilizing the properties of these media).

The laser head that Franken rented from Trion also figured in the development of a laser spectrograph. The story is interesting because it involved an active collaboration among Franken, who by June 1961 had become a consultant for Trion,[70] Trion staff members, and scientists at the instrument firm of Jarrell-Ash. It also shows how Franken's contribution emerged out of an interplay between his own research interests and Trion's market needs.

Franken had learned that Bell Laboratories researcher W. S. Boyle had used a laser to vaporize microscopic portions of carbon blocks. The plumes of vaporized material appeared to be self-luminous,[71] and from this fact Franken deduced that they were at temperatures of about 20,000°K. Franken reasoned that if the low-powered laser being used at Bell could create a plume of matter at 20,000°K, a higher-powered laser might deliver enough energy to reach the temperatures at which thermonuclear fusion reactions occur, so that the process might form a basis for fusion power or fusion weapons. He therefore began to repeat the Bell experiments in his own laboratory, using the Trion laser and his university's spectrograph.

The next stage occurred when the technical staff at Trion recognized that the phenomena Franken was studying might provide the basis for a laser technique of spectroscopic microanalysis. Trion's staff approached a number of instrument firms and eventually found an interested ear in Frederick Brech, vice president for research of Jarrell-Ash, a leading manufacturer of industrial spectroscopes. Jarrell-Ash and Trion scientists thereupon began a collaboration. Their early attempts to develop an instrument, however, were unsatisfactory. Even when the laser threw up adequate plumes, the spectral lines the plumes emitted were too weak.

Meanwhile Franken, pursuing fusion, discovered that the Bell scientists had been in error; the temperatures of the vaporized plumes were not 20,000° but only a few thousand degrees Kelvin. This explained the weak spectra. A laser could serve as a scoop to release the plume of material, but some other source of energy would be needed to excite the vapor to a point where it would fluoresce. Stimulated by a conversation with Brech, Franken and a colleague designed a setup using an electric field to provide the extra energy, thereby transforming the Jarrell-Ash/Trion laser spectroscope into a salable product.[72] Trion had provided a crucial piece of equipment for academic scientist Peter Franken, and as consultant, Franken contributed a crucial element to a Trion product.

By the time the work with Jarrell-Ash was well in hand, Trion was on its way out. Undercapitalized and overextended, it was no match for some of the stronger firms that were beginning to market ruby lasers. In July 1962, it sold itself to Lear-Siegler and became that company's laser arm.

The firm of Spectra-Physics, incorporated six months after Trion, was to have a happier destiny.[73] Perkin-Elmer, a leading manufacturer of precision instruments for scientific research and optical components for NASA and the Department of Defense, sought to coopt Spectra-Physics as its laser division even as Spectra-Physics was coming into being.[74] In the event, however, Spectra-Physics formed a joint venture with Perkin-Elmer that strengthened it during its infant years, and the company then went out on its own to become the largest U.S. producer of commercial lasers.

The five founders of Spectra-Physics had not started their firm expressly to produce lasers. They had broken away from the Stanford, California, instrument company of Varian Associates simply because they wanted to control their own destinies and build a firm that would be at the cutting edge of technology. The founders were physicist-turned-engineer Robert C. Rempel, Herbert M. Dwight, a Varian engineer with responsibility for selling R&D contracts to government agencies, the theoretician-experimentalist team of Arnold L. Bloom and W. Earl Bell, and systems engineer Kenneth A. Ruddock. They established their company in September 1961 to engage in research and development of electrical instruments, and especially quantum electronics instruments. They had no firm idea of the instruments they would make, beyond a general desire not to compete directly in Varian's market.

The five explored the venture capital markets but eventually decided to do their own capitalization, raising from among themselves and their friends a start-up sum of $225,000. Another important source of funding during the company's early years was R&D contracts, partly from local Bay Area firms but more importantly from government agencies. Like Technical Research Group in the 1950s, Spectra-Physics found that the government was an eager market. "In those days," Dwight reports, "It was relatively easy on an unsolicited proposal to go out and get a relatively nominal contract with somebody like the Naval Research Laboratory [NRL]. . . . A well-known researcher with a good idea could sit down with a top representative of the Naval Research Laboratory and pretty much on word-of-mouth commitment get money to do work . . . in promising areas." Some of the company's earliest contracts were with the NRL, the Air Force Cambridge Research Laboratories, and NASA. In its second year, it got a major, $580,000 contract from NASA. Total contracts reached $900,000 by May 1963.[75]

While they were getting started in fall of 1961, the founders had been approached by John Atwood, Perkin-Elmer's director of research. Atwood had heard through trade sources about the new company and wanted to acquire it. At this time, Perkin-Elmer was expanding into quantum electronics and was, in particular, developing plans to produce lasers.[76] Rempel, Dwight, Bloom, Bell, and Ruddock explained that they had no interest in leaving Varian merely to be swallowed up by Perkin-Elmer. In the course of their discussions with Atwood, however, they did arrive at a plan for a joint venture in lasers. The product they choose was the Javan-Bennett-Herriott helium-neon gas laser. They made this choice because Bloom and Bell were experts in gas discharge physics and had already worked on discharges with some of the same properties as the helium-neon laser, and also because AT&T could be expected to grant licensing rights under reasonable terms. The agreement between Spectra-Physics and Perkin-Elmer called for collaboration until the first 75 units were sold or two years had passed, whichever came first. Spectra-Physics was to provide the design and development work and Perkin-Elmer the optics, the manufacturing, and the marketing. The arrangement gave Spectra-Physics the use of an established name and a manufacturing and marketing capability, while it gave Perkin-Elmer the technical expertise it needed to enter the laser field. In March 1962, roughly three months after the arrangements were worked out, Perkin-Elmer/Spectra-Physics exhibited their first commercial laser at a

West Coast convention of the IEEE. The Model 100 produced an infrared beam at 1.15 microns and was priced just under $8,000.

The discovery at Bell Laboratories of lasing on a visible (red) helium-neon line in 1962 was critical to the new company's market success.[77] Earl Bell and Arnold Bloom's intensive study of the helium-neon discharge allowed them to pick up immediately on AT&T's results. The Model 110, introduced in September 1962, was the first visible helium-neon laser to appear on the market. Like Trion's ruby laser, the early Spectra-Physics/Perkin-Elmer helium-neon laser was far from perfected. The radiofrequency circuit used to excite the lasing gases interfered with other electronic devices being used in the laboratory.[78] Keeping the laser in alignment against mechanical and thermal perturbations was tricky. Nonetheless, a continuous, visible laser beam was almost irresistible to scientists and teachers. By June 1963, a half-year before schedule, Perkin-Elmer and Spectra-Physics had sold their 75th laser.

Spectra-Physics had grown restive during its year and a half of joint labor with Perkin-Elmer. To the little Silicon Valley firm, the Connecticut company appeared bureaucratic and cumbersome. Spectra-Physics fretted under delays in the delivery of the optical elements it needed for development work and in the manufacturing process. At Spectra-Physics, decisions were made quickly, mainly by President Rempel. The company, moreover, was on a high of energy and enthusiasm. Liberal stock options tied the employees to the company's future. Key scientific personnel were working six twelve-hour days a week and were often putting in time on Sundays.

During the period of collaboration, Spectra-Physics had added its own manufacturing capability and begun to make an increasing share of the product. It had acquired a marketing manager when it hired Eugene Watson away from Varian in late 1962. It had initiated research into the making of laser mirrors. When the 75th laser was sold, Spectra-Physics decided to go it alone. At this time, it was a firm of about 40 employees. A good percentage of its business was still government contract work, but Spectra-Physics was now committed to the commercial laser market for the long haul. For a while, Perkin-Elmer competed with Spectra-Physics for the laboratory market in helium-neon lasers, but ultimately it abandoned that market to its West Coast rival and focused instead on selling systems. Spectra-Physics sales went from $134,000 in its 1962 fiscal year to $1.7 million in 1964, and it was already able to show a profit in its 1963 books.

***

A company that hewed more closely to the classical scientist-founded model was Theodore Maiman's own firm, Korad. Maiman had left Hughes in 1961 to join Quantatron, a company just being started by a small group of scientists and committed to developing lasers as one of a number of high-technology products. In November 1962, Maiman struck out on his own with Korad, devoted wholly to lasers. The financing came as venture capital from Union Carbide, which retained 80% ownership. Maiman, and to a lesser extent a group of about ten cofounders, held the other 20%.[79] Union Carbide also had, however, an option to buy this 20% in stages, and after it exercised this option, over the ensuing five years, many of Korad's founders left. By the time Maiman left, at the end of 1967, the company had over 100 employees and sales of more than $5 million a year.

From the start, military orders and research contracts formed a large part of Korad's business. It was not easy for the military to know where to place its laser money in those early years. None of the new firms had had time to establish clear track records either for excellence or for slovenliness. Military monitors themselves did not yet know much about the science and technology for which they were contracting. TRG, which had received about $2 million in DoD money and had produced comparatively few discoveries, provided a cautionary example. Thus Maiman's proven success in inventing the ruby laser was a powerful attractant.[80]

Fred P. Burns, Korad's former manager of operations, recently estimated that about 50% of the company's budget went for R&D. This is to be compared with the 9.4% that was the average spent by segments of the electronics industry at the time and the less than 5% that was the average over all U.S. industry.[81] At Korad, as at every commercial laser firm, a large R&D effort was required to produce reliable products. The quality of the early ruby crystals, for example, was deplorable. Optical coatings blew up when struck by the high intensity of the laser beams. Techniques for producing high-power, fast pulses had been demonstrated in the laboratory, but the laboratory techniques had to be translated into forms adequate for commercial lasers. At Korad, coatings and crystal growing were done in-house. Integration was tempting to laser firms because there was as yet no industrial infrastructure, no network of businesses producing the high-quality crystals, glasses, and optical elements that lasers required. Korad had active research programs in gas lasers and semiconductor lasers, as well as in ruby and other solid-state devices. In part, this reflected a desire to develop new products. In part, however, it came about because many of the scientist-founders came from Hughes Research Laboratories and were accustomed to research of a more basic nature than was usual, or prudent, in small firms. One Korad researcher recalled, "We did lots of exciting, innovative research. The sense of competition was so great that we would run to the mailbox to get *Applied Physics Letters* as it arrived. Our attitude was that although we were small, we were going to show that we could do good science."[82]

In generalizing from this triad of cases, we see, first, that small firms contributed their own share of fundamental results to the laser boom. Bloom, Bell, and Rempel at Spectra-Physics were the first to publish on the high-gain 3.39-micron line in helium-neon, and Earl Bell marked out a major new direction when he invented the mercury ion laser at the end of 1963. Small firms made advances in the technique known as "Q-switching" and published on a variety of new laser media.[83] They were also responsible for a number of successful laser applications. We have already looked at the case of Trion and the laser spectroscope. Starting in 1963, Spectra-Physics put up considerable money of its own, in a collaboration with another firm, to develop an instrument, called the Geodolite, for making profiles of the height of terrain or of ocean waves.[84] Korad carried out a large research program on laser welding under a contract with Linde, a Union Carbide subsidiary.[85]

These companies also contributed to the laser boom by providing off-the-shelf lasers to laboratories in every sector of the research establishment. To be sure, these early lasers often worked poorly. They had in some cases to be redesigned, or even set aside in favor of homemade lasers constructed to the users' specifications. But even at worst they were a means by which a new laser scientist could get his or her feet wet.

Turning to the question of what motivated and permitted these little companies to enter the field, we must begin by noting that there was a widely spread sense that the laser market would take off quickly. In 1962, a New York firm, Technology Markets, Inc., predicted that 1970 sales would reach $1 billion, 75% of that to the military.[86]

There was also capital from a variety of sources. Trion and Spectra-Physics capitalized themselves from the savings of the founders and money from small individual investors. Korad turned to the corporate surplus of a large conglomerate. A prominent early Boston firm, Maser Optics, was started when a New York investment banker put together a private group of affluent investors who wanted to get into the laser business.[87] Optics Technology, a West Coast firm, was financed by an investment company owned by the Bank of America, as well as by several private investors.[88]

Barriers to entry were low. It was not difficult for physicists or engineers to put together a laser in the laboratory, once they had learned something of the technology from Maiman or the Javan group. The limited size of the market meant that production could be carried out by scientists or skilled technicians, without a need to redesign the product so that it could be assembled by cheaper, less skilled labor.

Nor were patent rights a barrier to entry. To begin with, the patent situation was uncertain in the years through 1963. Charles Townes had received a patent for the maser in March 1959, and he and the nonprofit Research Corporation, to which he had assigned the rights, took the position that the maser patent also covered optical masers. Two objections, however, were being raised by the legal staffs of some laser-selling firms. One was simply that the scope of the maser patent did not extend to the optical region; the other was that Townes's patent was invalid because he had "published" his results in the Columbia Radiation Laboratory Quarterly Reports too far in advance of filing his application. The Research Corporation was to succeed in securing its first licensing agreement (with AT&T) only in 1965.[89]

In March 1960, Townes and Schawlow had been issued a second patent, assigned to AT&T, on the optical maser. But that patent, which clearly did cover lasers, was soon embroiled in litigation. TRG, Inc., brought an interference before the Patent Office on the grounds that Gordon Gould had had prior conception.[90] The other patent applications—for example, those on Javan's helium-neon laser and on Hughes's lasers—were still pending.[91]

Laser patents were the less a problem in that AT&T claimed some of the most important of them. AT&T occupied a unique position. In 1956, as the outcome of a long process that started when the Truman administration decided to increase competition in the electrical manufacturing industry, AT&T signed a consent decree that allowed it to keep its manufacturing arm, Western Electric, but imposed a number of restrictions. The relevant one here is that AT&T was required to grant nonexclusive licenses to its patents on terms whose reasonableness was to be guaranteed by court oversight.[92] Laser firms could devote their energies to developing inventions from Bell Laboratories into good products, knowing that they would be able to negotiate for licenses with modest royalties when AT&T eventually came after them.

It is also pertinent that a considerable proportion of the lasers and laser systems were being sold to the government. In many cases, the government had rights to royalty-free use; where it did not, prosecution for patent infringement was too difficult to be worth attempting. Sales to academic research projects supported by government sponsors were also exempt. Finally, the diminutive size of the laser industry itself conferred protection. Lawyers do not usually take companies to court unless the sales that will be covered under a license agreement are large enough to yield royalties that cover costs. The small laser market thus made litigation uninteresting.[93] Major laboratories such as Bell and Hughes therefore functioned as R&D resources for the newly emerging laser firms. It was a symbiotic relationship rather than a parasitic one, however, for industrial scientists, as well as academic ones, were profiting from the off-the-shelf equipment the commercial firms were beginning to provide.

Laser Research in the Defense Industry

At the large defense firms, the focus was on systems. Production of commercial lasers, when it occurred, was a sideline. Hughes Aircraft Company and Raytheon Corporation will serve as our examples in this section. Large companies also provided the locales within which we can best explore the interactions between systems research and device research and between applied research and more basic research. I shall make use of two episodes in the history of the Hughes laser program to illustrate these two characteristic aspects of laser history.

At Hughes Aircraft, lasers already claimed company attention as "Invented Here." In addition, management was attracted to lasers because it viewed them as having enormous potential for both military and civilian markets. The company was at the time largely tied to a single buyer, the U.S. Air Force, but its traditional product, aircraft electronics, was being challenged by the rise of missiles. Management wanted to diversify. While they sought to retain a strong share in the military electronics market, they also hoped to build a commercial business large enough to account for perhaps a third to a half of the company's sales within a decade. The laser could play an important role in each of these target markets. For this reason, Hughes had a companywide laser committee and laser coordinator as early as September 1960. From two lone investigators, Maiman and D'Haenens, in early 1960, the number of scientists working on lasers grew to nine by September 1960. In 1963, Hughes had a total of 55 staff members working on laser projects.[94]

Hughes worked on several laser systems in the early 1960s, including systems for communications, metrology, and beam weapons. Its major focus, however, was laser radar, and here it quickly became an industry leader. Radar was an obvious application.[95] It was universally recognized that, were it possible to advance from microwave to optical radar, the accuracy with which a target could be located and distinguished from other targets could be greatly improved. Whereas conventional radar might locate an enemy to within 900 feet at 50,000 feet, laser radar could locate an object to within a foot.[96] The antennas for optical radar would also be a fraction of the size of microwave antennas, an advantage in mobile systems such as tanks, planes, and space vehicles. Radar systems using incoherent light had already been attempted, but because their light beams lacked the directionality and monochromaticity that would ultimately be furnished by laser light, they were usable only at short range and in the absence of sunlight.

From Hughes's viewpoint, laser radar had a number of advantages. It was one of the most immediately realizable of laser applications, its use of pulsed signals meshed with the properties of the laser Maiman had just developed, and the cost of development would probably be supported by Department of Defense contracts.[97] Laser rangefinders also conformed to the Hughes management principle of selecting for development systems that were both technologically difficult, so as to cut down on competition, and likely to develop major military markets.

By the end of 1960, Hughes had two contracts for laser radar programs. One, from the Air Force, was for work by George F. Smith's Exploratory Physics Department at the Research Laboratories at Malibu. The other, from the U.S. Army's Frankford Arsenal, was for a project under Malcolm L. Stitch at the Culver City facility in metropolitan Los Angeles. Smith's and Stitch's groups both succeeded in establishing the feasibility of laser ranging with breadboard apparatuses by the end of January 1961 (figure 4.1).[98]

Commercial ruby laser production was an offshoot of the radar work. It was overseen by Douglas A. Buddenhagen, the member of Smith's team who had built the ruby laser for the rangefinding experiment. Buddenhagen had gone on to develop smaller and more reliable versions of the laser, under contract to DoD groups and for research use within the laboratory, before management chose him to organize a commercial unit. Hughes here was following an already established pattern of taking devices that were of interest to the company as part of its systems and also selling them as individual components. A previous example had been microwave tubes. It was a type of diversification that helped to smooth out company revenues.[99] The company hoped for appreciable sales.

Figure 4.1
Hughes engineer Rod Smith prepares to fire the Colidar laser rangefinder. Laser light, emitted from the long barrel, is reflected back from the target to the telescope in the shorter barrel. A timing circuit measures the elapsed time for the light beam's round trip and computes the distance. (Courtesy Hughes Aircraft Company and AIP Center for History of Physics)

"In 10 or 15 years," executive vice president Roy E. Wendahl explained, "[lasers] will play an important role in everyone's life and standard of living."[100] On a technical level, management expected that laser improvements made by the commercial group and the systems groups would feed into each other.

Buddenhagen's group, which numbered 7–10 members at its maximum, constituted a minuscule entrepreneurial firm within Hughes.[101] Its rate of production of ruby lasers—priced at $2,500, including the power supply—ran at about 20 per month. The group also produced custom-designed units on special order. The first model was put on the market in May 1962, and by mid-1963 sales were about $100,000 per year. Even Buddenhagen's group, charged as it was with producing a component, made an effort to develop an application, in this case a laser welder. But merely to develop a marketable pulsed ruby laser consumed most of the group's effort. The same trio of problems with which Trion and Korad struggled—securing high-quality ruby crystals, optical coatings that resisted damage, and long-lived flash lamps—preoccupied them also.[102]

The Raytheon Corporation offers another example of how commercial production could fit as one piece of a mosaic of laser activities within a large defense firm. Unlike Hughes, Raytheon got into lasers only after Maiman announced his results in July 1960. Raytheon was then a diversified electronics company with more than $500 million dollars of sales per year. About 80% of its products were for the military.[103] In 1960, scientists and engineers at giant facilities and diminutive companies alike were going to their laboratory benches and using the published articles, their own expertise, and the know-how informally communicated by others in the field to recreate the ruby laser. This is what scientists at Raytheon did. Once they had a Maiman-type laser working, two independent groups, one at Spencer Laboratory, which housed the Microwave and Power Tube Division, and the other at the Research Division a few miles away, went on to construct a version that used a linear flash lamp in place of the spiral one Maiman had employed. To maximize the transfer of power from lamp to ruby, Raytheon placed the lamp on one focal line of an elliptical reflector and the crystal on the other. With this geometry, the pump power could be cut down to one-fifteenth of that required from the spiral lamp and the ruby could be more easily cooled.[104] This was an important advantage because the time required to cool the ruby between shots was severely limiting the number of pulses per second that could be got out of the laser.

Meanwhile, Raytheon customers were asking the Research Division how they could obtain a laser. Thomas H. Johnson, division director, therefore took the elliptical ruby laser to Howard Scharfman, manager of the Special Microwave Devices Operation (SMDO). This unit had recently been formed to help Raytheon rework its technological innovations into marketable products. SMDO was closely connected with the Research Division. It was initially housed in the same building, and it was common for research personnel to move over to SMDO to work on the development of Research Division ideas. Circumventing the usual time-consuming procedure of detailed market research, Scharfman and SMDO Technical Manager Colin Bowness instituted a crash program. As a result, Raytheon was able to exhibit what was probably the first commercial laser, to be sold for research purposes, at the March 1961 New York City meeting of the Institute of Radio Engineers. By August 1962, the SMDO commercial catalog listed several ruby lasers, including a high-powered variant that had been used by MIT in its June 1962 Luna See experiment, in which laser light was reflected from the moon. SMDO also sold a helium-neon laser. Management saw in the SMDO laser program a means not only of developing marketable lasers but also of creating proprietary knowledge that could make Raytheon's military systems more competitive.[105]

By late 1962, a small but vigorous laser group had been built up within the Research Division. It was under the direction of Hermann Statz, who served at once as laser theoretician, enthusiast, and promoter of laser systems research within the company as a whole. A structure under which Statz had become one of a triumvirate of managers of the division permitted him scope. The group took virtually all the extant lasers—crystal, gas, and semiconductor—under its purview.[106] It published equally on theory and experiments designed to reveal the detailed behavior of lasers. In the physics journals, where publication criteria dictated that only scientific advances be reported, the Research Division's results appear as a kind of torsoless head, yet this work was strongly nourished by technological considerations. For example, the Surface Radar and Navigation Operation of Raytheon's Equipment Division had an active program in laser radar, as well as a program in radiation weapons, by the end of 1962.[107] Statz foresaw using the laser for Doppler radar, which reveals the velocity of a moving target as well as its distance. But Doppler radar requires the transmitter to radiate a single frequency. This was one of the factors behind the investigation that Chung Liang Tang, Statz, and George A. de Mars published in June 1963 on simultaneous oscillations at multiple frequencies in solid-state lasers. To understand how the multiplicity of modes arose was a step in creating the base of knowledge on which single-frequency lasers could be erected.[108]

By the end of 1962, SMDO was projecting a budget of $750,000 for 1963.[109] Yet neither at Raytheon nor at Hughes did the production and sale of lasers per se ever loom large in comparison with the work on laser-based systems. Indeed, none of our evidence contradicts the general pattern in the electronics industry that large firms tend to concentrate on systems and smaller firms on components.

It is worth noting that while newly established firms such as Trion and Spectra-Physics labored to develop systems that could provide a market for their lasers, the reverse was often true at the large companies. That is, researchers at the larger companies were often already at work on systems for which the laser provided an alternate component. American Optical's laser photocoagulator is a case in point. Photocoagulation had been developed in the late 1940s as a therapy for detached retinas. The procedure used the light of a xenon arc lamp, which was directed through the eye's pupil and lens to a spot on the detached portion of the retina, where the light energy heated the retinal tissue and welded it back in place.[110] At the American Optical Company, where ophthalmic instruments were already a part of the product line, research physicist Charles J. Koester was working on a photocoagulator utilizing a mercury arc when Elias Snitzer suggested that he explore the ruby laser as the light source. Koester developed a laser instrument with the help of ophthalmologist-consultant Dr. Charles J. Campbell. Koester subsequently delivered an improved version to Campbell's laboratory in the Institute of Ophthalmology at the Columbia-Presbyterian Hospital in Manhattan. In December 1961, the Columbia-Presbyterian group used a laser on a human patient for the first time, destroying a retinal tumor with the American Optical photocoagulator.[111] Campbell and his colleagues also used the laser in their ongoing exhaustive examination of optical photocoagulating agents. Meanwhile, American Optical started marketing laser photocoagulators at the modest rate of a dozen or so a year.[112]

Hughes Aircraft Company's 1961 and 1962 radar work provides a good example of the intersection of systems research and research to improve lasers. The radar research done by the Culver City and Malibu groups in late 1960 and early 1961 helped to underscore a major weakness of the ruby laser. The onset of the laser pulse was unpredictable and uncontrollable, and when it finally came, the output was not a single, short, well-shaped pulse but a series of irregular spikes, stretching over many milliseconds. This meant that reading the range out of optical radar results was cumbersome: Instead of simply measuring the time that elapsed between sending a pulse out and receiving a reflected signal from the target, one had to compare the incoming train of spikes with the outgoing one and determine the offset between the two patterns. Even more serious was the length of the train of spikes. Ambient air particles would be still scattering portions of the outgoing light back into the receiver as the reflected train arrived, so that the reflected signal was bathed in noise.[113]

Radar was only one of a number of laser applications being handicapped by spiking.[114] The attempts made to explain the phenomenon, and the still greater number of attempts to control it, were the more important for systems research in that the ruby was, at the time, one of a very small number of available lasers. The Hughes staff participated actively in this work. At the Culver City plant, for example, Stitch's group made a direct attack on spiking in an effort to improve the laser radar they called COLIDAR (for Coherent Light Detection and Ranging).[115]

Still more important was the work on spiking done at the Research Laboratories at Malibu. Robert W. Hellwarth wrote a major theoretical paper on the phenomenon in late 1960,[116] and he drew upon this theoretical work at the Second International Quantum Electronics meeting in March 1961 to suggest a dramatic improvement in the laser. Hellwarth predicted that a single spike of giant power could be created if the reflectivity of the laser's end mirrors were suddenly switched from a value that was too low to permit lasing to one that was sufficient, a technique that would ultimately be called Q-switching. The switching could be effected in several ways; Hellwarth suggested inserting a Kerr cell—a device that rotates the plane of polarized light when a voltage is applied to it—together with a polarizer in front of one of the laser mirrors.[117] He calculated that peak powers as high as 12 megawatts might be achieved in this way. This figure was thousands of times higher than the power of Maiman's first ruby laser and a billion times larger than laser power estimates made by Schawlow and Townes a mere three years earlier. Soon after, experimentalist Fred McClung, who had been working in Smith's rangefinder group, joined forces with Hellwarth to bring a Q-switching system into operation. In their early experiments, they got peak powers of 600 kilowatts, about 100 times those of ordinary ruby lasers.[118] Hughes tried immediately to fit Q-switching by Kerr cells into their radar systems.[119]

One classic instance of the interaction between basic and applied research grew directly out of Q-switching. The work of Hellwarth and McClung meant that Hughes was, for the time, one of the few laboratories that had high-powered Q-switched lasers. This helped bring in contracts. Malcolm Stitch's Laser Development Group at the Culver City Laboratory obtained an Air Force contract in the spring of 1962 to study the optimum conditions under which high-power light from a Q-switched ruby laser could be further amplified by means of a second ruby laser. Eric Woodbury was in charge of the program, and he was joined by a colleague, William K. Ng. They used two different types of detectors to measure the light from their Q-switched primary laser. They found that when they operated the ruby laser without Q-switching, both detectors gave the same result, but when they activated the Q-switching, the detector readings were inconsistent. Woodbury and Ng provisionally tracked the effect to the nitrobenzene-filled Kerr cell that was part of the Q-switching mechanism and shared the resonant cavity with the ruby rod. The nitrobenzene appeared to be the seat of several new radiations that had lower frequency—infrared as against the ruby's red line—but otherwise exhibited the properties of laser light. The new lines were narrowly collimated, had sharp widths, and only came into being when the ruby laser light reached a certain threshold output power. Woodbury and Ng were able to show that the two types of detectors they were using were equally sensitive to red ruby light but differently sensitive to infrared light and that that had been the reason for the inconsistent measurements.[120] In July 1962, Woodbury and Ng sent a letter describing their findings to the IEEE *Proceedings*; they were well aware, however, that they had not yet solved the scientific puzzle of how the new lines were being generated.[121]

At Malibu, Fred McClung got wind of the result before it was published, through a visiting scientist, Edward Condon, who had stopped at Culver City before arriving at Malibu. McClung had also been troubled with discrepant results from his monitoring detectors and had consulted Woodbury on the problem. The situation displayed some classical elements of the tension between the urge to compete in science and the urge to cooperate. Woodbury and Ng knew that they had uncovered a new phenomenon, and now they had something of a proprietary interest in the discovery. McClung, for his part, had a proprietary interest in Kerr-cell Q-switching and a laboratory well equipped to tackle the further steps.

The situation also reflected social factors arising from differences in the roles played by the scientists at Malibu and Culver City. McClung and his colleagues in Smith's Exploratory Devices Department had considerable freedom, within broad boundaries, to choose their own research directions. They were working under an Air Force contract that essentially directed them to study the laser in whatever way they thought profitable. They were caught up in the general scientific excitement of lasers and in the group spirit of the laser work at Malibu, which was characterized by informal competition coupled with a hearty appreciation of other people's achievements.[122]

Woodbury and Ng, in contrast, were under pressure to fulfill an Air Force contract that focused on the amplifying laser rather than the Q-switched laser that supplied the signal to be amplified. Further, Stitch's Culver City operation was primarily charged with bringing new system concepts to the stage of well-engineered, marketable products. Like every other manager at Hughes in this financially precarious period, Stitch was also interested in bringing in contract money, and a contract was in fact negotiated that enabled Culver City investigators to pursue the study of the new effect in other materials. But overall, the ball passed from Culver City to Malibu. Other Malibu staff, including, independently, Gisela Eckhardt and Robert Hellwarth, eventually succeeded in identifying the effect as a new species of laser emission, "stimulated Raman scattering." Ordinary Raman scattering was well known; it involved the transition of a microscopic quantum system from a low-energy state to a higher-energy state through the simultaneous absorption and emission of two photons of differing frequencies. The existence of a lasing version of this process was significant because it simplified the creation of population inversions.

Stimulated Raman scattering was a basic discovery, but it emerged out of an applied project, the ramping up of the power output of ruby lasers. It shows how even the applied laser investigations being undertaken by scientists and engineers were contributing to the discovery of fundamental effects. Phenomena lay about like nuggets in an unexploited gold field, and the amount of discovery was bound to be proportional to the sheer amount of effort applied. There is no doubt that Woodbury and Ng's work on amplifiers accelerated the discovery of stimulated Raman lasers, but by how much is hard to say. For at least one deliberate effort to produce stimulated Raman scattering had been initiated—by Herbert J. Zeiger and Peter E. Tannenwald at MIT's Lincoln Laboratory—at the time that the Hughes scientists discovered it by accident.[123]

The First Gas Lasers

The mood of excitement that drove many U.S. researchers in 1961 and 1962 was evident in the work on gas lasers. Advances occurred in four broad directions. The laser resonators were improved, for example, with more conveniently configured mirrors. New methods of excitation were discovered. New gases were made to lase. Finally, spectroscopic studies led to an enhanced theoretical knowledge of the behavior of the lasing substances. These branches of work cross-fertilized each other. The more the resonators for existing lasers were improved, the easier it became to make new gases lase, for the technological advances in mirrors and tubes relaxed the conditions needed to secure oscillation. New methods for exciting population inversion, invented either in the course of empirical work or as a by-product of spectroscopic studies, also opened the door to additional classes of lasers. The synergism enhanced the pace of discovery. It was the very fruitfulness of laser research, the rapidity with which interesting results could be attained, that constituted one of the principal causes of the excitement.

The laser researchers of 1958–1960 had in the main worked slowly, carrying out extensive calculations and careful preliminary experiments as a prelude to testing their devices. But with the proliferation of new types of lasers, the uncertainties these pioneers had felt began to fade from the collective memory. It began to seem as though almost anything could be made to lase, and this in itself was another incentive to try new media. In an article in spring 1963, a *Fortune* writer quoted one scientist as saying, "I expect any day now to hear that someone has got a tube of plain air to lase."[124]

Bell Telephone Laboratories, which had the largest overall laser program in America at the time, was a major site for gas laser research. *Barron's* speculated in February 1963 that Bell was spending $5 million a year on laser research. The other major industrial programs, such as those at Hughes or Raytheon, were probably about one-fifth as large. Of the 106 papers listed in *Physics Abstracts* for 1962 and 1963 as having at least one of its authors situated at a U.S. industrial laboratory, 67 were from Bell.[125] Javan, Bennett, Herriott, and Schawlow had worked within departments in the Physical Research Laboratory, and J. H. Sanders had worked as part of James Gordon's group in Rudolf Kompfner's Electronics and Radio Research department. Starting in 1961, laser research spread rapidly to other Bell departments. P. K. Tien's microwave tube department reoriented itself to gas laser work. The technology that underlay gas lasers, such as glassblowing and high-vacuum techniques, was similar enough to microwave tube technology to make conversion rapid. Scientists already in the group redirected their research, and Tien also hired new people specifically committed to laser work, including Herwig Kogelnik, an Austrian physicist-engineer, and C. K. N. Patel, an Indian national trained at Stanford. Because the laboratories were expanding, Tien could hire freely, while laser research was attractive to exactly the sort of highly qualified scientists Bell Laboratories sought. Departments in the development area as well as the research area gravitated toward lasers. H. E. D. Scovil's solid-state maser department within the Solid-State Device Laboratory started laser research, as did members of groups within the Electron Tube Division.[126]

One immediate achievement at Bell was the first operation of a helium-neon laser with external spherical mirrors. Rudolf Kompfner suggested this project, and it was carried out by William W. Rigrod and Herwig Kogelnik from Tien's group in collaboration with Herriott and senior mechanic D. J. Brangaccio. The original helium-neon laser of Javan, Bennett, and Herriott had plane parallel mirrors inside the laser tube. These mirrors were easily damaged in the course of the heat treatment that constituted one step in tube construction. They were further damaged by the electric discharge within the lasing gas. They also needed exquisite alignment: A deviation from parallelism of a few seconds of arc was enough to quench the laser.[127]

The means to overcome these disadvantages were at hand even before December 13, 1960, the day on which the helium-neon laser first operated. Work on spherical mirrors at Bell Laboratories dated back to the spring of 1960, stimulated by the informal laser seminar that met at Bell's Murray Hill facility. At that seminar, Fox and Li had presented a numerical solution for resonators with both plane and curved mirrors, done with the aid of a computer, and there had been some grumbling among auditors about the absence of an analytical solution. W. Deming Lewis, the director of communications systems research, had remarked that an analytical solution would be possible if two identical spherical mirrors were placed at each other's focal point (confocal mirrors). This led a recently arrived member of the Bell technical staff, Gary D. Boyd, who had already been experimenting with curved-mirror resonators, to start working with Lewis and with James Gordon, his supervisor, to formulate a mathematical treatment of confocal resonators. Spherical mirrors relaxed requirements for parallelism by almost two orders of magnitude. Boyd and Gordon also showed that they concentrated the light around the axis of the tube.[128] This would open the way to smaller-diameter tubes, which is important because the gain that can be achieved in a gas laser increases approximately inversely with diameter.

The alternative to the internal mirrors that Javan, Bennett, and Herriott used would have been to lead the light out of the discharge tube, through windows, to a pair of external mirrors. Some of the light would inevitably be lost passing through the window. It was a well-known piece of optical science that the loss could be minimized by orienting the windows at a particular angle, the so-called Brewster angle (figure 4.2). The Javan team elected not to experiment with Brewster angle windows. "We needed every last ounce of gain," Bennett explained later.[129]

It was a different matter once the helium-neon laser had been made to work. Rigrod and his collaborators used the strongest of the lines that Javan's team had discovered to demonstrate that the

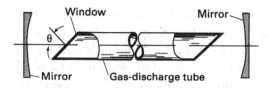

Figure 4.2
External mirrors and windows oriented at the Brewster angle θ.

losses in the Brewster windows would not extinguish the laser oscillation. The use of external mirrors eased both construction and operation of the helium-neon laser. From a piece of apparatus that had to be babied, it became sturdy and portable enough to be used for applications research.[130] Moreover, this type of resonator became a valuable instrument for studying new gas lasers. Bennett pointed out that "the use of external mirrors . . . permits rapid change of mirrors for covering different wavelength regions, and also makes it easier to deal with chemically active gases," while the fact that the spherical mirrors allowed "using long lengths of small-diameter tubing for initial studies of new systems has been quite important."[131]

The helium-pumped cesium vapor laser, successfully operated at TRG in the early months of 1962, was one of the lasers that the new resonator structure made possible. Helium-pumped cesium had been glancingly mentioned in Gordon Gould's proposal and patent application as possible but not promising.[132] Not surprisingly, therefore, TRG's first experiments with alkali vapors made use of potassium, the lasing medium that had been suggested by Schawlow and Townes. The choice reflected the close collegial relations that existed between Stephen F. Jacobs, head of the TRG experimental team, and Herman Z. Cummins, the senior of the two graduate students that Townes had assigned to the Columbia University project.[133]

The Columbia connection was also one of the reasons why Jacobs and his junior colleague, Paul Rabinowitz, switched from potassium to cesium sometime in the first half of 1960. Oliver Heavens, who was assisting the Columbia laser work during the spring 1960 term, had given a seminar at TRG at which he had forcefully argued for cesium.[134] Gould also argued for a switch to cesium, animated by a deep sense of frustration that none of the TRG experiments had yet yielded a laser.[135]

The work at Columbia University had changed direction after 1960. Oliver Heavens had returned to England. Cummins had continued working with cesium even after Maiman's announcement, in the hope that he could come up, if not with the first of all lasers, at least with the first continuous-wave maser. Javan, Bennett, and Herriott copped that honor in December 1960, however, and Cummins phased out his cesium work and turned to other problems.[136] A cesium laser held no particular interest except as a "first." Like the helium-neon laser, it promised low power, on the order of a milliwatt. In addition, its probable operating wavelengths would be so far into the near infrared as to pose difficulties in finding detectors and transmitting materials, and it would be an intrinsically difficult device to construct and use because of cesium's extreme reactivity.

At TRG, however, other considerations obtained. The company had received over $1 million in DoD contracts and was under pressure to show its contract monitors some results. The recollections of company president Lawrence Goldmuntz may well represent emotions at the time. "When Maiman's laser came out, it was after our contract was signed, we began to get a little bit nervous. We'd gotten the bulk of government money. We were doing good work. But we hadn't gotten the laser to work. So we felt that it would be important for us to show that we were able to reduce one medium of the many that we had suggested in our proposal to practice, and it was felt that [cesium] would be the clearest one that one could do. . . . I think our calculations showed it was not likely that was going to be a very powerful laser, or very meaningful laser or anything like that. But it was an attempt to do something in a medium that was pretty controllable, and where you could understand some of the processes more readily than you could with some of the other approaches. That I think was the primary motivation for doing it."[137]

Jacobs and Rabinowitz thus continued their efforts, in consultation with Gould. By March 1961, they had demonstrated unequivocally the existence of inverted populations for the 3.20-micron and 7.18-micron transitions in cesium vapor. By September 1961, they had obtained laser amplification on these lines, using a meter-long gas tube. They succeeded in obtaining oscillation on the 7.18-micron line in early 1962. An arrangement of near-confocal mirrors, one flat and one spherical, eased the problem of alignment for this technically difficult laser. Brewster windows allowed them to protect the mirrors from cesium, and a cleverly designed vacuum system geometry made possible a cesium-free environment for the vacuum seals.[138]

Javan left Bell Telephone Laboratories for a faculty position at MIT at the end of the summer of 1961. Bennett was now joined by two younger men, Walter L. Faust and Ross A. McFarlane. Bennett, whose graduate training and early work were in spectroscopy and who knew well the literature and the experimental techniques of that field, moved continually between his personae as spectroscopist and as device-maker, deducing from his laser experiments new spectroscopic facts or puzzles in need of explanation, and conversely deriving from the results of spectroscopy new ideas for lasers. His style, like Javan's, was to find out every relevant physical fact it was possible to know before attempting to make a system lase.[139]

One of Bennett's concerns was the creation of new types of lasers. Even before the first helium-neon laser was made to operate, he and Javan had made the accidental discovery that excited neon atoms seemed able to transfer enough energy to the oxygen molecules that were present as a trace impurity in helium-neon mixtures to dissociate the molecules, leaving oxygen atoms in excited states. They postulated that it might be possible to make an oxygen-atom laser by using a mixture of neon and molecular oxygen. This would have important advantages. The helium-neon laser works because the helium atom in one of its excited states has just the right energy, within a few tenths of an electron volt, to raise a neon atom from its ground state into an upper laser state. The class of pairs of atoms with such exact energy coincidences, however, is tiny. In contrast, the match that is required between an excited atom and a ground-state molecule in order for the atom to transfer its energy to the molecule is 10 times less stringent. "Dissociative excitation transfer" therefore appeared capable of generating a considerable number of new lasers.

Bennett was also interested in the possibility of pure-neon lasers. He and Javan had originally decided that they were impossible, because their experiments indicated that the lower laser level would be too heavily populated by secondary processes. Subsequently, however, Bennett reassessed the relative strengths of the several processes, and in early 1962 he, Faust, and McFarlane were able to show that they could in fact get population inversion in pure neon.[140] In the absence of helium, neon was being excited by the impact of energetic electrons. Electron impact was an even more general excitation method than molecular dissociation by excited atoms, and it opened the way to still another host of new lasing systems.

Before Bennett and his collaborators could move from a demonstration of inversion to laser action, quite independently another Bell scientist, C. K. N. Patel from Tien's group, made pure neon lase (figure 4.3). Patel, unlike Bennett, had only a nodding acquaintance with spectroscopy when he arrived at Bell Laboratories in June 1961. He was just 22 years old, and his degree was in electrical engineering. He therefore took a more empirical path when he approached atomic spectroscopy. Patel knew that Javan and Bennett often chose to keep their laser gases pure by evacuating the laser tube, filling it with gases, and then disconnecting the tubes from the vacuum pump and sealing it off to prevent contamination. Patel, who in any event was not a skilled enough glassblower to seal

Figure 4.3
C. Kumar N. Patel at his Bell Telephone Laboratories workroom in 1966 with a high-energy carbon dioxide laser. (Courtesy C. K. N. Patel)

off his own tubes, elected to keep the pump on the tube continually and to flow his gases through the tube in order to flush out contaminants. "A very small idea," Patel later explained, "but it turned out to be very useful, because . . . changing gases in a tube becomes a . . . trivial matter." Rather than embarking on lengthy calculations, the young engineer decided to make a parametric study of the action of the helium-neon laser as the ratio of helium to neon was varied. "One of the first things I did was I started cutting back on the helium pressure, and surprise of all surprises, you didn't need helium to make the neon laser work."[141]

Bennett, Patel, Faust, and McFarlane began to cooperate at this point, purchasing mirrors jointly, coordinating their experiments, and publishing together. They put the neon-oxygen system into operation and then extended that genre of devices by making an argon-oxygen molecular-dissociation laser. They created lasers in pure argon, xenon, and krypton, as well as finding new lasing lines in previously explored systems such as neon and helium-neon. They altered their laser tubes to make them capable of sustaining oscillation in the farther reaches of the infrared and showed that the wavelength region longer than 2 microns was rich with laser transitions. Patel, Faust, and McFarlane continued the work after Bennett returned to Yale in the fall of 1962. "These were the days when if a day went by when either myself or Faust [and] McFarlane didn't find 10 new lines . . . there was something basically wrong with us."[142] By early 1963, they had achieved lasing on about 150 additional lines through various excitation mechanisms. The work demonstrated that, in gas discharges at least, population inversions were more the rule than the exception.[143] The copiousness of the discoveries they were making sustained their enthusiasm. "There was not a single day that passed when something really exciting didn't happen here. And I think for that reason [my wife] put up with late nights . . . day-in and day-out, all Saturdays, all Sundays."[144]

What Bennett was to call "the 'prettiest' results in the helium-neon system"[145] came out of the electronic components development area within Bell Laboratories. There, in an exploratory development group within the Microwave Tube Department, two device scientists, Alan D. White and J. Dane Rigden, had a small contract from the Signal Corps laboratory at Fort Monmouth to build a helium-neon laser with a direct-current gas discharge. Direct-current excitation seemed likely to be more efficient and more easily controllable than the radiofrequency excitation Javan and his collaborators had used. In the course of this work, Rigden and White came to the conclusion that the neon line at 6328 Å in the red part of the visible spectrum might also be made to lase through an excitation process distinct from, but closely analogous to, the process by which Javan and his collaborators had produced their infrared laser lines. This would give the continuous-wave gas lasers an advantage heretofore accruing only to the pulsed ruby laser; visible light was easier to detect than infrared and made it easier to study laser properties. In addition, it appeared that visible light would be easier to modulate, and thus a laser that was both continuous-wave and visible might be useful as a transmitter for optical communications.[146] Work in the device area was more strongly oriented to practical application than was work in the research area at Bell Laboratories,[147] and White and Rigden pursued the visible helium-neon laser on weekends and evenings while completing the Signal Corps contract during working hours.

Initial measurements made with the type of discharge tube they were using for the Signal Corps project, which had an internal diameter of 15 millimeters, failed to demonstrate gain on the red line.[148] Nevertheless, Rigden and White sent the Bausch and Lomb optical company an order for special mirrors that would be highly reflecting in the region around 6328 Å. They also decided to change their 15-millimeter tube for one of 7 millimeters, since they expected that gain would increase as the diameter decreased.

Rigden and White did not have complete spectroscopic information. They did not, for example, have an estimate of the lifetime of their upper laser level. There was no time for careful preliminary research in their schedule. White later recalled, "We were aware of the tremendous amount of work Javan, Bennett, and Herriott had done on the [infrared] laser. . . . We couldn't do that kind of work in evenings and weekends even if we had [had] the equipment. The simplest thing to do was to set up the tube, align the mirrors, and try it. The use of Brewster windows and spherical mirrors simplified our work immensely. . . . [Afterwards] Bennett said something to the effect that there were two ways to confirm the possibility of inversion on a potential laser line. One was to make a methodical and detailed spectroscopic study. The other was to use an analog computer consisting of a gas discharge tube and two mirrors. And in fact, within six months a great many people were using the second method."[149]

Clearly, gas laser research was not driven solely by the exuberance of young workers attacking a novel field. In particular, the wholehearted support that managements gave these projects was motivated by an interest in eventual applications and markets, and sometimes by the immediate prospect of R&D contracts. Nevertheless, there was an unmistakable aura of "new lines for new lines' sake" about this early period. This was particularly true for gas lasers as opposed to solid-state lasers, for the low output powers and clumsy apparatuses of the gas devices made their practical employment appear more remote. In the meantime, the desire to advance science, the wish to advance one's own career through the publication of a stream of new results, the spur of intrainstitutional competition, and to a considerable extent sheer enthusiasm provided the incentives.

Semiconductor Lasers

One measure of the excitement the laser generated is the extent to which members of scientific specialties outside the community of quantum electronics became involved in the search for new lasers. Perhaps the most important were the semiconductor scientists. The semiconductor transistor, invented in 1947, had created a research field so vigorous that by the mid-1950s semiconductors commanded about 1,000 publications annually.[150] With such a large pool of scientists and so many material specimens available for experimentation, it is not surprising that many teams of semiconductor experts plunged into laser research. In the United States alone, in 1961 and 1962 work was undertaken at RCA, Lincoln Laboratory, IBM, General Telephone and Electronics, and General Electric to build a semiconductor laser. A race developed that was to end in a photo finish in the fall of 1962, as General Electric, IBM, and Lincoln Laboratory all independently crossed the finish line within a month of each other.

Even before Maiman's ruby laser, there had been proposals for making semiconductor lasers.[151] Electrons in pure semiconductors can occupy energy levels in the "valence band" or higher levels in the "conduction band"; between these bands is a zone of forbidden energies. Pierre Aigrain of the École Normale Supérieure in Paris suggested using the transition of electrons from the conduction band of silicon or germanium into the valence band, with the emission of a photon of light and the simultaneous release of a quantum of energy to the thermal vibration of the crystal lattice, as the basis for a laser.[152]

At Lincoln Laboratory, Benjamin Lax, inspired by a talk Aigrain gave at MIT in 1957, started exploring transitions among a group of energy levels created when a semiconductor is subjected to a strong magnetic field; these transitions would give rise to photons in the millimeter and infrared range.[153] Jacques I. Pankove of RCA Laboratories spent a fellowship year in 1956–1957 working with Aigrain in Paris. He learned of Aigrain's ideas in November 1956, and in early 1957 he collaborated in their theoretical and experimental elaboration.[154] In Japan, Yasushi Watanabe and Jun-ichi Nishizawa filed a patent application in 1957 for tellurium and silicon semiconductors masers that they expected would be capable of producing radiation in the near infrared. At the Lebedev Institute in Moscow, Nikolai Basov, in collaboration with B. M. Vul and Yu. M. Popov, started to work in 1957 on a proposal to use impurity-doped semiconductors. The impurities add energy levels within the forbidden zone to the electron's options. Basov and his collaborators suggested using both transitions between the bands and the impurity levels and transitions between the bands themselves. And at the beginning of 1960, William S. Boyle and David G. Thomas of Bell Laboratories filed a patent application that discussed the use of both elemental (silicon and germanium) and compound (such as gallium arsenide) semiconductors as lasers and suggested several excitation methods.[155]

To this point, in 1960, there was a wide area of uncertainty. No one knew whether a semiconductor laser was possible. Even if it were, it was not clear that any of the schemes proposed so far would work. There were questions about the kind of semiconductor that should be used—whether it should be elemental or compound, whether it should be pure or doped, whether it should be a homogeneous crystal or the junction formed between two halves of a semiconductor crystal doped with different kinds of impurities. It was not clear whether the resonant cavity should be placed around the semiconductor, as in the solid-state maser, or whether the crystal could be made into its own cavity (as Maiman was doing with the ruby laser by silvering the ruby's two ends). No one knew which of the several proposed methods for pumping the laser would work.[156] Nor was there as yet an unimpeachable derivation of the conditions that would have to be met to reach the threshold of lasing.

Then, in 1960, Maiman's work proved that lasers were possible. The subsequent work at IBM, Bell Laboratories, and Varian proved that lasing was not just a freak property of pink ruby. This had an effect on the search for a semiconductor laser. In 1961 and 1962, in addition to the work at Lincoln Laboratory and RCA, research was inaugurated at IBM, General Telephone and Electronics, and then General Electric.

At IBM's Watson Research Center, Rolf W. Landauer had just moved from a position as head of a small theory group to manager of a section comprising both basic semiconductor research and lasers. This gave him a natural role as an initiator of semiconductor laser research, and he began to talk up the topic among his scientists and to his Army Signal Corps contract monitors.[157] He was particularly attracted by the possibility of injection pumping, which would obviate the need for the unwieldy array of optical lamps then universally used for pumping solid-state lasers. Such a simplification might pave the way for a continuous solid-state laser, which was then one of the objectives of the IBM laser group. In December 1961, Landauer established a small study group to examine the problem in a more systematic fashion.[158]

Some months earlier, a representative of the Army Signal Corps Laboratories at Fort Monmouth had visited the Watson Center to hear Landauer and his group present their ideas on semiconductor injection lasers. At the end of the year, the Fort Monmouth Procurement Office issued a request for proposals leading to a working, experimental, injection-pumped semiconductor laser. IBM submitted a proposal and was subsequently awarded the contract. The approach IBM stressed used the semiconductor crystal as a host for impurity atoms whose energy levels would be prominently involved in the lasing transition.[159] Meanwhile, William P. Dumke, who belonged to another section of the laboratory and was not affiliated with Landauer's study group, performed an analysis leading to the conclusion that the whole class of indirect semiconductors, to which silicon and germanium belonged, would be unsuited for lasing.[160] Dumke also pinpointed the direct semiconductors, especially gallium arsenide, as promising materials. The area of uncertainty about the kind of material that would be required thus began to contract. IBM was well placed to investigate indirect semiconductors because it had an active program at this time in the preparation and study of gallium arsenide semiconductors.

Compound semiconductors, and especially gallium arsenide, were also being studied at the Bayside, New York, facilities of the General Telephone and Electronics Laboratories (GT&E). There Sumner Mayburg was leading a group of about ten scientists and technicians exploring what kinds of devices might profitably be made from gallium arsenide. This required them to deepen their knowledge of gallium arsenide's properties, and among the diagnostic techniques they employed was a study of the radiation produced at a junction during injection when an electron in the conduction band recombined with a hole in the valence band (recombination radiation). Toward the end of 1961, Harry Lockwood and San-Mei Ku in Mayburg's group each independently made a discovery implying that under certain circumstances virtually every charge sent across the junction gave rise to a photon corresponding to the band-to-band transition. This was a finding akin to Maiman's determination, two years earlier, that almost every photon pumped into a ruby crystal showed up as a photon in the lasing transition.[161] It suggested that pn junctions would be preferable to homogeneous crystals for a laser and that injection pumping would be effective.

By this time, the possibility of a semiconductor laser had become a major topic at semiconductor conferences. Mayburg himself had been speculating on how to do it. He and his scientists had access to specialists since they daily rubbed elbows with GT&E's active gas and glass laser groups. The discovery of the high efficiency of conversion of current into band-to-band radiation led Mayburg to redirect part of his group's effort into laser research. Unlike IBM, the GT&E group gave a minor role to the investigation of semiconductors as a host for impurity atoms. It concentrated instead on the direct band-to-band transition in compound semiconductors.[162]

Pankove at RCA also submitted a proposal to Fort Monmouth.[163] There had been a small semiconductor laser effort at RCA Laboratories since Pankove had returned from France. He had not, however, been able to interest his management in a stepped-up program. Laboratory director William Webster estimated that the probable return to the corporation from the sale of laser products would not be large enough to warrant the investment.[164] At the January 1962 meeting of the American Physical Society, however, Pankove reported that he and his colleague M. J. Massoulie had made measurements on direct band-to-band recombination radiation excited by injection in heavily doped gallium arsenide junctions.[165] Mayburg now saw that he might be scooped by RCA, and he hurried to get his group's results into print.[166]

Sumner Mayburg and Rolf Landauer had been college roommates at Harvard,[167] and in early 1962 Landauer invited Mayburg to present his laboratory's results to IBM technical staff members. Mayburg did not mention lasers to the IBM scientists, but he did report Ku's result of intense gallium arsenide luminescence. He also gave qualitative arguments to support 100% efficiency for the radiative recombination process. The effect of this lecture at IBM was galvanic. Gordon Lasher, a theorist and member of Landauer's study group, sat down to try to work out a resonant cavity structure for a semiconductor laser. He turned up a major uncertainty. It was impossible to predict whether the structure he designed would focus or defocus the laser light; if, and only if, it focused it, lasing looked possible.[168] Marshall I. Nathan, a semiconductor experimentalist who was studying radiation from gallium arsenide but who to that point had not given much attention to Landauer's laser program, now became directly involved. Drawing comfort from Lasher's results, he began to devote his odd hours to making a gallium arsenide injection laser.[169]

An essential precondition for Nathan's informal search was the strength of the IBM program in gallium arsenide. For several years, a group under Richard Rutz's direction had been looking into technological applications of gallium arsenide diodes. In the course of this work, they had mastered techniques for creating the tiny pn junctions and for fixing metallic contacts to them to permit the flow of current. They had learned to make junctions of very small area, a skill that was to prove crucial to getting the high values of current density (current per unit area) that would be needed.[170] Nathan could obtain the samples he needed from Frederick Dill of Rutz's group, without placing significant demands on the group's time or resources.

At Lincoln Laboratory's Solid State Division, a group under Robert H. Rediker had been studying gallium arsenide diodes since 1958.[171] In mid-1961, this group began to look at radiative recombination in order to explicate differences between diodes formed by two different fabrication techniques, diffusion and alloying. The experiments showed the same startlingly strong luminescence that GT&E and RCA were discovering. Rediker writes, "The . . . luminescence from the diffused diode pinned the recorder, and the servomechanism that actuated the recorder was churning. [Robert J.] Keyes [who was conducting the experiment] had to increase the full-scale setting of the recorder by at least 3 orders of magnitude and close the spectrometer slits to near zero to bring the reading back on scale."

The Lincoln Laboratory team reported its results in July 1962 at the Solid State Device Research Conference at the University of New Hampshire. Pankove of RCA had presented similar findings at a meeting of the Electrochemical Society in May. Now the entire community of workers on gallium arsenide semiconductors was privy to the news of gallium arsenide's remarkable luminescent efficiency. The results even made it into the pages of the *New York Times*, because MIT scientists had used the luminescence to transmit a television signal 275 feet.[172] Nathan had not attended the conference, but he recalled that the publicity went as far as top IBM management and that their interest was transmitted back as a pressure for an IBM contribution to this newsworthy field. Nathan now perceived that he was in a race with Lincoln Laboratory and RCA.[173] In truth, the race was wider still.

One entrant was a group in General Electric's Schenectady laboratory. GE semiconductor physicist Robert N. Hall had participated in conference conversations on the possibility of semiconductor lasers.[174] He had been interested enough to inform himself about lasers. In particular, he had taken advantage of the yearly visit of French scientist and semiconductor laser theorist Maurice G. A. Bernard. In late 1961, Bernard and Georges Duraffourg had published an analysis of the requirements for lasing in semiconductors.[175] But Hall had not been sufficiently optimistic about the outcome to start a research project. He was jolted out of his pessimism by the New Hampshire conference. Hall recalled in a 1984 interview that the high intensity of radiation reported by Lincoln Laboratory and RCA "was really astonishing. Previously we thought maybe 0.01% efficiency might be reasonable, and here they are talking close to 100%. So that really shook me up; that was the lightbulb turning on."[176] On the train returning to Schenectady, he started sketching a scheme he thought might work.[177] Hall's background in gallium arsenide junctions, recombination radiation, and the optical properties of semiconductors gave him the knowledge that he needed for this work.

A few days later, after he had mapped out preliminary plans, Hall recruited a team of part-time collaborators. There was Ted Soltys, an experienced fabricator of semiconductor devices, Jack Kingsley, an optics and laser expert, and semiconductor scientists Richard Carlson and Gunther Fenner. He also secured his manager's approval. "The contract load was relatively modest and there weren't many project-oriented programs, so it was fairly easy to rearrange priorities."[178] It was important that the gallium arsenide and ancillary equipment were already in the stockrooms. This meant that the only cost to the company was the part-time salary of the scientists.[179] It also meant there would be no loss of time waiting for apparatus to arrive. "It was evident that we had to work fast. Every other semiconductor lab in the country knew about this GaAs luminescence and at least two of them, Lincoln Laboratory and RCA, had a head start of several months."[180] Hall organized a division of labor among his team members, with Carlson supervising the doping of the crystals, Soltys fabricating tiny diodes with polished edges out of the doped material, Fenner testing the diodes for lasing, and Kingsley helping to analyze the results (figure 4.4).

The New Hampshire conference also inspired Nick Holonyak, at General Electric's Syracuse laboratory. Holonyak belonged to the community of workers on gallium arsenide and had, in fact, for several years shared with Hall an Air Force contract on gallium arsenide research and devices. Holonyak decided to try making a laser of a gallium-arsenic-phosphorus semiconductor that would lase on a red line. This would make it possible to detect the onset of lasing through the change in the visible emission, a far easier procedure than using instrument measurements on the infrared light from gallium arsenide junctions. The semiconductor he chose was an exotic; there were no diodes conveniently on the shelf. Holonyak, however, was a pioneer in these materials and an expert in their fabrication. In common with all the others, though, Holonyak faced the problem of how to situate the diode within a resonant cavity. Hall had decided to use the diode as its own Fabry-Perot cavity, cutting two of the faces perpendicular to the plane of the junction and parallel to each other and polishing them to make mirrors. Holonyak first inclined toward placing the diode within an external cavity; later he settled on the strategy of trying to cleave the semiconductor material along its natural planes and using these cleavage planes as mirrors.[181]

Figure 4.4
Robert N. Hall with his semiconductor laser, here suspended inside a glass
container filled with liquid nitrogen. In the reflecting prism below the glass
container, the laser crystal can be seen as a small spot between two connecting electrodes. (Courtesy General Electric Research and Development Center
and AIP Center for History of Physics)

Rediker and his colleagues at Lincoln Laboratory were also trying to follow up their results by making gallium arsenide lase. They had gone to see Herbert Zeiger, who drew from his own earlier calculations suggestions for the geometry they might use. Solid State Division head Benjamin Lax, like Hall's manager at General Electric and Landauer at IBM, gave this project his vigorous support.[182]

At IBM, Nathan consulted Peter Sorokin about lasers and talked to Dumke and Lasher about the problem of how to detect lasing if it did occur. After a number of failed attempts, Nathan, on Dumke's advice, decided to look for an abrupt narrowing of the spectral linewidth, which would mark the transition from ordinary light to highly monochromatic laser light. On September 28, 1962, Nathan raised the current through a pn junction in stages from 0.1 milliampere to 25 amperes and saw the linewidth contract by a factor of more than four (figure 4.5). "At this point, I stopped. I went over to [Dumke's] office to show him the results. We were both ecstatic. . . . Over the weekend, Gerry Burns and I went to the lab to confirm the result and to see how narrow a line we could get."[183] The IBM results were hustled through the patent office and sent by messenger in October to *Applied Physics Letters*. Shortly thereafter, to Nathan's surprise, he discovered that not RCA or Lincoln laboratory, but General Electric, had crossed the wire ahead of him.[184] The article by Nathan and his collaborators had arrived at the journal on October 6, but Hall and his team had gotten results to the *Physical Review Letters* on September 24. Holonyak, after he learned of Hall's success, gave up his attempts at cleaving and polished mirrors onto his diodes. His group assembled its first polished-cavity $GaAs_{1-x}P_x$ laser on October 9. On October 10, using Hall's equipment, Holonyak and his coworkers demonstrated lasing. At Lincoln Laboratory, Rediker's group lased gallium arsenide during the first part of October; their results reached *Applied Physics Letters* on October 23. Mayburg's group at GT&E only succeeded in late November, after the other papers were in print and after obtaining a specially designed power supply to provide the high current densities they needed.[185]

It is remarkable that the first semiconductor lasers were operated in the United States, when so much of the early work had been done in other nations. How is this to be explained? One factor was the sheer size of the American semiconductor industry, then by far the

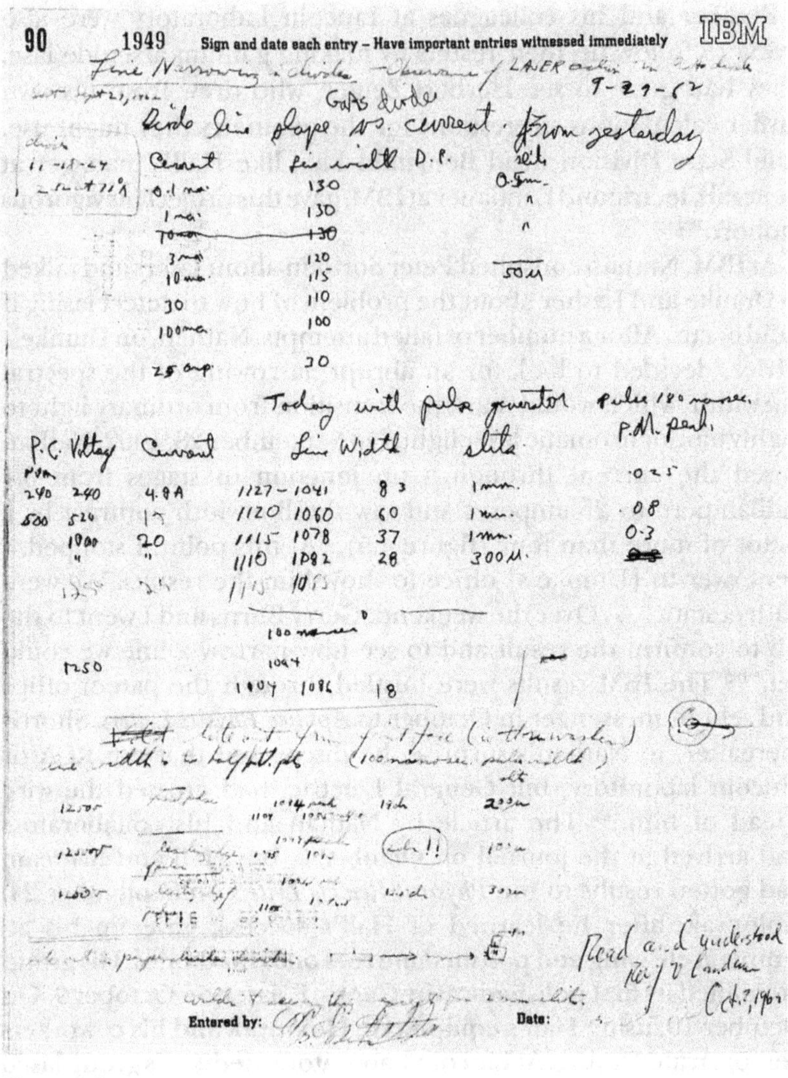

Figure 4.5
This page from Marshall I. Nathan's notebook shows his excitement as he drove the current density through the semiconductor diode up past the threshold of laser action. (Courtesy M. I. Nathan)

largest in the world,[186] and the attendant magnitude of U.S. semiconductor research. Almost without exception, the researchers discussed above were working on compound semiconductors; they were familiar with their properties and skilled in their fabrication. Once the excitement engendered by the discovery of the laser had spilled over into the semiconductor world, this expertise became a formidable resource for laser-making.

The large stock of semiconductor materials and equipment that was on hand was of crucial importance, as we have seen. It meant that the work at IBM, General Electric, and the other laboratories could be done on a part-time basis, with a relatively small commitment of resources. In this connection, we must recall that the semiconductor laser was a high-risk project. Robert Hall had estimated his chance of success as 1 in 5.[187]

A second factor that may help explain U.S. priority is the spur competition gave to the work. The judgment that other groups were on the trail, and a felt need to hurry, was marked at each of the winning laboratories. Competition came into play particularly after the New Hampshire conference in July 1962. There was also some fear of foreign competitors. At IBM, there was a rumor in early 1961 that Pierre Aigrain was coming to the March meeting of the Optical Society of America with a semiconductor laser in his pocket. Holonyak recently recalled that he was chiefly worried about competition from the Services Electronics Research Laboratory in Great Britain.[188] But it appears to have been domestic competition that most inspired the feverish pace of work.

The semiconductor injection lasers announced at the end of 1962 created a sensation. They had exceptionally high efficiencies; by the end of 1964, these reached as high as 70–80%. They were tiny, typically 0.5 millimeter long and 0.1 millimeter in lateral dimension, as opposed to the centimeter lengths of ruby crystals and the more than a meter size of gas lasers. They were excited by applying an electric current to the semiconductor, whereas all the other solid-state lasers made to this time had used optical pumping. As a consequence, they could be easily and rapidly modulated, since merely varying the strength of the current would vary the intensity of the beam. *Electronics* headlined an early article, "Will Diode Laser Obsolete Earlier Lasers?"[189]

The first application for semiconductor lasers on everyone's lips was optical communications. Modulation of the light from gas and solid-state lasers was difficult, and the ease and rapidity with which the amplitude of the semiconductor beam could be modulated was very attractive. A second use that soon entered discussions was optical computing. This was, perhaps, inevitable. The U.S. computing industry was undergoing an extraordinary expansion. Installed computer capacity was increasing at a rate of 65% a year. The business grew from virtually zero in 1955 to about $2 billion in 1961 and more than $17 billion in 1965. Many new companies were entering the market.[190] Furthermore, computer technology was unusually plastic. The commercial introduction of transistorized computers, the so-called second-generation computers, had occurred from 1958 through 1960, but it was by no means clear to all research scientists that transistors would remain the best way to go for computer logic.[191]

The Air Force's Rome Air Development Command had already been funding the study of lasers for computing, with one contract at RCA and another at American Optical to study the use of glass fiber lasers. Glass lasers, however, had been shown to require too much pumping energy.[192] Gallium arsenide injection lasers were several orders of magnitude more efficient than glass lasers, and this drastically ameliorated the pumping problem. They were small and were rapidly being reduced to dimensions comparable to the transistor. The semiconductor laser appeared to be a candidate for computer logic, for computer peripherals, and also for communication among computer subsystems, and a lively program in these areas took shape at IBM and elsewhere after 1963.

The Laser in the Popular Press

Excitement within the research community found an echo in the popular press. *Time* magazine included the laser in its cover story of January 2, 1961. This was its traditional "man-of-the-year" issue, but in 1961 *Time* chose instead to focus on U.S. science. The cover portrayed 15 scientists, including Charles Townes. In sober, well-researched articles, *Time* traced the progress of U.S. science from a dependence upon Europe to a position as a recognized world leader. Its thumbnail biography of Townes included his development of the radiofrequency maser from 1951 to 1954 and carried the story through to the operation of optical masers in 1960. The maser, radiofrequency and optical, *Time* judged as "one of the most revolutionary devices of the age . . . of immense practical application not only on earth but in seeking out the wonders of the universe." "In solid state physics, the maser replaced the transistor [in 1960] as the hottest of all items."[193]

Lasers were the subject of feature articles in other wide-circulation magazines. *U.S. News* ran a story headlined "'Light Ray'—Fantastic Weapon of the Future" in April 1962 and another, "Miracle Ray is Coming of Age" in February 1963; *Life* did a photo-essay on the "Amazing Laser" in January 1963; *The Reader's Digest* commissioned an original story, "Light of Hope—or Terror?" in February 1963; *The New York Times Magazine* published "Laser Lights Up the Future" in September 1963; and the *Saturday Evening Post* came in with "The Astounding Laser" in October 1964.[194]

All of these articles portrayed the laser as a discovery of the first importance. The transistor and the vacuum tube were frequently cited analogies. All listed a number of potential applications, with the ones chosen for mention well-correlated with the interests of the scientists they cited as informants.[195] Yet two applications were given universal attention. One was beam weapons. The transformation of the Buck Rogers ray gun into a battlefield reality appears to have been irresistible to reporters and editors. The lead paragraph of the *U.S. News* story is typical of much of the reporting: "Suddenly the idea of 'death rays'—light shot through guns—has moved out of the realm of science fiction." *The Reader's Digest* subheadline was: "The present and potential uses of the laser—a new kind of light ray—sound like science fiction. In fact the invention is one of the most amazing accomplishments of our time." And the magazine later explained more specifically, "Many experts predict that someday the laser's 'bullets of light' may usher in a kind of warfare so far envisaged only in science fiction."

The other attention getter was an experiment carried out at MIT by engineers Louis Smullin and Giorgio Fiocco in the spring of 1962, in which they beamed the light of a ruby laser at the moon and recorded the reflected echo. This work was an extension to lasers of the maser radar astronomy done earlier on, and it was grounded in the same mix of scientific and military interests. The MIT experiment represented two technical feats. The first was the construction of the most powerful ruby laser yet built, the Raytheon 50-joule laser. The second was the accomplishment of the MIT scientists in fishing the return signal from the moon, which their initial estimates showed would be about 12.5 photons per pulse, out of the background light, which they had estimated at 30–80 photons per pulse. The *New York Times* gave this event first-page coverage under the heading "Man Shines a Light on the Moon." *Life* featured it in its January 1963 photo-essay, and *The Reader's Digest* began its article by describing it. In June 1963, *Fortune* dubbed it the most famous of the early stunts.[196]

Above and beyond the difference in emphasis given to the various applications by the scientific community and by journalists was a still more fundamental dissimilarity. Scientists were excited by the laser because the properties of coherence and monochromaticity that were characteristic of the radio region were now being carried up to the optical region, leading to the possibility that light might be made to perform all the technological functions of radio waves. This possibility was largely ignored by the popular articles. Instead, they invariably introduced the laser as a means of transforming light into an exalted state. Thus, "regimented light unlike anything known before" (*Reader's Digest*), "strange new light" (*Time*), "ordinary light . . . [turned] into a beam of incredible sharpness and power" (*U.S. News*, describing optically pumped lasers), and, from *The New York Times Magazine*, "light stops being just something to see by. It becomes a powerful tool." This was, of course, a way of explaining the new discovery to the laity. But it is hard to escape the impression that the press's presentation had as its unexpressed and perhaps unapprehended basis the magic and power that has long been associated with light in the human imagination.

In December 1964, Charles Townes, Nikolai G. Basov, and Alexandr M. Prokhorov received the Nobel prize in physics for fundamental work in quantum electronics leading to maser and laser devices (figure 4.6). The Nobel prize is an accolade bestowed by a group within the scientific community upon selected members of that community. In light of their earlier articles, however, the popular publications that reported the award had every reason to expect their readers to appreciate why the laser, at least, was worthy of the honor.

Figure 4.6
Alexandr M. Prokhorov (left) and Nikolai G. Basov (right) host Charles
Townes in their Moscow laboratory in September 1965. (Courtesy N. G.
Basov, A. H. Guenther, and AIP Center for History of Physics)

5

Out of the Laboratory and into the Marketplace

Introduction

Around 1964, laser euphoria began to show signs of wearing thin. Some of the projected applications were proving noticeably difficult to achieve, for a variety of reasons. In some cases, the lasers themselves were recalcitrant. It was, for example, turning out to be unexpectedly hard to develop a room-temperature continuous-wave semiconductor laser for computing or communications purposes, or a solid-state laser that was sufficiently powerful for use in beam weapons systems. In other cases, it was the auxiliary elements for laser systems that posed the problems—the modulators and demodulators needed for optical communications systems, for example, or the transmission lines needed for terrestrial optical communications. Sometimes unexpected side effects appeared, as when scientists attempting to use the laser to treat cancer discovered that the blasts of light were scattering malignant cells from the tumor to other sites. At other times, more detailed analyses revealed that the traditional technologies simply had imposing physical and economic advantages, as, for example, the transistor increasingly appeared to have, vis-à-vis the laser, for computer logic.

A measure of disillusion set in. Some of the organizations that had initiated early programs drew back. Westinghouse, Hughes, Raytheon, and American Optical scaled back or even eliminated some of their commercial or research programs. The Advanced Research Projects Agency within the Department of Defense curtailed its laser beam weapon program.[1]

The equation of the laser with the transistor, so common in the first years of the 1960s, began to be superceded by a contrast. The laser, it began to be said, was unlike the transistor because the transistor could simply be plugged in wherever vacuum tubes had previously been used but the laser did not lend itself to such simple substitution. A new catchword now spread. The laser was "a solution in search of a problem." That is to say, the true market for the laser would be in applications that were as yet undiscovered and unexpected.[2] Clearly, this was only partly true. The laser was certainly sufficiently radical an invention that it was bound to have uses that were not immediately foreseen. But there were also plenty of "plug-in" uses for it—in place of xenon arc lamps in photocoagulators, for example, or in place of mercury arcs in interferometers. Moreover, the lasers of 1964, with their short lifetimes, their low efficiencies, and the spottiness of their coverage of the electromagnetic spectrum from near infrared through visible to near ultraviolet, could scarcely be qualified as full-fledged "solutions." Nevertheless, while the slogan did not accurately capture the technical situation, it did represent the sense of frustration within the laser industry at the slowness with which the market was developing.[3]

Even as a more sober mood settled in, however, the research results continued to pour out. The number of laser-related papers listed in *Physics Abstracts* were over 1,000 per year by the late 1960s.[4] This research advanced laser science and technology along a great diversity of paths.

New types of lasers were invented. Carbon dioxide and other molecular gas lasers joined the older types using excited atoms. Lasers with ionized rather than neutral atoms were discovered; the most important among these was the argon ion laser. Vapors of copper and other metals were made to lase. Liquid lasers made of quasiorganic molecules dissolved in solvents were invented, and after these came the dye lasers, an important class of wholly organic lasers. New types of crystal hosts took their places beside the earlier sapphire (aluminum oxide, which, when doped with chromium, makes the ruby crystal) and calcium fluoride. One of the most significant of these was yttrium aluminum garnet, known as YAG. New methods of exciting laser media entered the armamentarium of the scientists. One of the major innovations was excitation through chemical reaction; another was excitation by rapid cooling through expansion, using processes taken over from the field of gas dynamics.

Within the field of gas lasers, there were a variety of improvements in the resonator structures, including better ways to glue the Brewster windows to the tubes, more optimal tube dimensions and tube shapes, tube materials that were less porous and more resistant to damage, and longer-lived electrodes. The early radiofrequency-excited helium-neon lasers, for example, had lifetimes of only a few hundred hours before helium diffused out through the walls of the silica-glass tubes. By replacing the silica-glass with Pyrex, and the radiofrequency excitation by direct-current excitation from hot cathodes, developers increased lifetimes to 1,000 hours. The hot cathode was the new limiting factor. In 1966, a cold aluminum cathode was introduced, and after some engineering, lifetimes increased to 5,000 hours.[5]

In the field of solid lasers, materials research led to laser glasses that were freer of contaminants. Better techniques for crystal growing produced larger host crystals and more careful control over the active ions with which they were doped.

A good deal of work was done to shape the output beam. Most gas lasers oscillate at the same time in many independent modes, each with its own frequency; the number of oscillating modes increases with the length of the resonator. Single-mode gas lasers were created by the simple expedient of using very short resonators, but also by the more sophisticated procedure of designing mirrors whose reflectivity varied so sensitively with frequency that only a single (or at most a few) mode(s) could surmount the lasing threshold.[6]

"Mode-locking" was invented. With this technique, a periodic disturbance introduced into the laser cavity enforces a fixed relation among the phases of the modes. The character of the output beam depends upon the relation among the phases, which in turn depends upon the wavelength of the disturbance. Mode-locking was first applied to helium-neon lasers, and a succession of regular pulses, each of a few nanoseconds duration, were produced (a nanosecond is 10^{-9} second). The technique really took off, however, when it was applied to solid-state lasers. United Aircraft Corporation's Research Laboratories was at the forefront of this research. There, Anthony J. DeMaria and his colleagues mode-locked neodymium-doped glass lasers, getting continuous trains of pulses with thousands of megawatts peak power and durations in the range of picoseconds (a picosecond is 10^{-12} second). These pulses were a thousand times shorter than any signals generated by scientists to that date. Sylvania and Stanford University scientists, also by means of a mode-locking process, were able to create a "supermode" from a gas laser, a single mode containing almost all the energy of the original independent modes.[7]

Theorists developed deeper and more detailed models of laser behavior. The earliest theories had been based on rate equations, which expressed the net rate at which the active systems were fed into the upper laser level (or the lower one) as a result of all possible processes of excitation and relaxation (loss of energy). Such theories were now succeeded by models, formulated by Willis E. Lamb jr. and others, that took into account the dynamic interaction of the lasing system with the electromagnetic field. First came theories that treated the lasing atoms as quantum-mechanical systems but the field as a classical system; these were followed eventually by completely quantum-mechanical theories.

These and other developments increasingly expanded the laser's range of possible applications. The continuous-wave carbon dioxide laser, as it was driven to higher and higher powers in the second half of the 1960s, revived the Department of Defense's interest in laser beam weapons. It also opened up the possibility of rapid cutting of nonmetallic materials such as cloth. The argon ion laser proved to be superior to the ruby for retinal photocoagulation. Ultrashort pulses from mode-locked lasers permitted technical applications such as high-speed photography and scientific applications such as the study of physical and chemical phenomena so short-lived that they had previously been unobservable. Dye lasers, because they could be tuned over greater ranges of frequency than previous lasers, promised to make it possible for the first time to do spectroscopy in the visible region comparable to that done with klystrons in the microwave region. These technical advances were also creating new niches within the laser industry that would allow for new market entrants with substantially different products. Two counterbalancing moods were therefore reflected in the literature of the mid-to-late 1960s. One was a sense of disappointment or pessimism in the face of the slowness of the development of the laser market and laser technology. The other was a sense of excitement and enthusiasm over the pace of the scientific advances and the commercial and technical possibilities they brought into being.

Mood and technical advance were not the sole determinants of the process by which lasers were transferred into the marketplace. The changing economic, social, and political environment also played a role. The détente between the United States and the Soviet Union that brought a decline in spending for strategic weapons in the mid-1960s has to be placed alongside the difficulties in achieving adequate high-energy lasers in explaining the dip in the military's interest in directed beam weapons. The growing American commitment to the Vietnam war accelerated work on tactical weapons such as laser radars, target designators, and reconnaissance systems. Concern for the environment and, later, for energy led to the release of federal and private money for exploring applications such as laser air-pollution monitoring and laser energy technologies. Thus the movement of laser systems from laboratory to workplace in the United States reflected the interaction of the constantly improving technical capabilities of lasers with a constantly changing pattern of societal demands.

Different applications were affected by differing combinations of these forces and therefore moved forward at different rates. Tactical weapons, pushed by government R&D spending and procurement, formed one of the earliest markets to mature. Materials processing, a field in which there were many competing technologies and little DoD interest, lagged. The development of lasers into industrial systems for alignment and measurement was comparatively straightforward technologically, and these applications blossomed early. Energy production through laser fusion received much interest and money but was so technically horrendous that even today it has not been demonstrated as a practical process.

The evolution of the laser industry was also governed by a mixture of economic and technical factors. The national economic environment grew harsher in the decade starting in 1964. The research market, a principal source of revenue for the commercial laser sector, contracted as funding for R&D declined, in constant dollars, from 1968 to the mid-1970s.[8] Capital also became difficult to obtain in this period.[9] Yet technical progress enabled the laser industry to hold its own and even grow. The inaugural market report of *Laser Focus*—the first American trade magazine devoted exclusively to the laser industry, and itself both a symbol and promoter of the industry's expansion—reported in January 1966 that the number of firms manufacturing lasers had gone from less than 20 at the beginning of 1963 to about 115 at the end of 1965.[10] The Electronic Industries Association estimated that the total revenue of the industry, including R&D contracts, lasers, and systems incorporating lasers, increased rapidly after 1965, reaching about $90 million in 1969. A *Laser Focus* review in 1971 indicated a continuation of this trend: It estimated $120 million for 1969 and $138 million for 1970.[11] Manufacturers' associations began to emerge. Spurred above all by the threat of government safety legislation, and spearheaded by the editor of *Laser Focus*, William Bushor, laser companies and government, industry, and academic scientists formed the Laser Industries Association (LIA) in January 1968. In response, the already established Electronic Industries Association (EIA) set up its own rival Laser Subdivision.[12]

From 1964 on, the U.S. popular and trade press took to proclaiming that lasers had graduated from "laboratory curiosities" into practical devices. *Missiles and Rockets* headlined a 1964 article, "Laser Emphasis Shifting from Research Toward Application and Hardware: Speed of Technological Development Gratifies Experts; Big Payoff Expected in Surveillance and Reconnaissance, Weaponry, Space Communications, and Data Processing." *Electronics* later that same year gave its readers an article entitled "Laser Welders: out of the laboratory and into the factory." "It looks now like the laser is ready to graduate from the lab," explained the reporter.[13] These proclamations may tell us something about how editors go about attracting readers, or about the habits of thought of journalists or of their sources among scientists and public relations personnel. They say little about the actual rate and process of commercialization. In reality, it took more than a decade of work on the science and engineering of lasers, of development of laser systems, and of growth and solidification of the laser industry, before lasers moved in any numbers out of the laboratory and into the marketplace.

Technical Advances: The Creation of New Lasers

I shall follow a single thread, the invention of novel laser types, from among the many lines of technical advance in the period 1964–1970. Although it leaves to the side many interesting developments, this focus will illuminate a number of important issues. It will allow us to see how technical progress opened the doors to additional applications. It will also show how new technology created new niches for the formation of companies. In addition, I shall argue that it demonstrates that work on laser devices was increasingly constrained by the demands of laser systems. More specifically, I shall claim that the search for new lasers was more influenced by application requirements in the mid-1960s than it had been either in the era of creation, 1958–1960, or in the ebullient years that followed in 1961–1963.

The creation of new laser types was an obvious strategy for moving forward. The lasers that had been operated through early 1963 showed a wide variety of properties. Their light varied in frequency from the red of ruby and helium-neon down to infrared wavelengths near 30 microns. Their peak power varied from the hundreds of megawatts of Q-switched ruby and neodymium-doped glass to the milliwatts of gas lasers. They differed in the total energy of their output, the temperature at which they functioned, their method of pumping, the efficiency with which they converted pump energy into output energy, the degree of angular divergence of their emitted beam, and their size (including both the laser head, comprising laser tube and flash lamps, if any, and the power supply). It stood to reason that one potent method for achieving still other properties was to create new lasers.

A steady stream of papers attested to the employment of this strategy. If we look at the cumulative index to the 1965–1968 *Engineering Abstracts* and count the number of distinct laser media mentioned, we find roughly 100. There are about 50 neutral or ionized gases listed, among them various halogens, noble gases, noble gas ions, and vapors of metals such as copper, zinc, and cadmium. There are a variety of molecules, including carbon dioxide, carbon monoxide, water vapor, nitrogen, and hydrogen cyanide. There are more than 20 different semiconductors, from the original gallium arsenide through binary compounds of cadmium, zinc, and lead and on to ternary compounds such as gallium arsenic phosphide and cadmium sulfur selenide. There are about 10 different crystals and 5 glasses.

Fortunately for the historian, who would be hard put to treat the evolution of 100 new lasers, work was highly concentrated around the one or two most promising types within each category. The number of entries dealing with neodymium-doped glass outnumbers the total of all other glass laser researches by about 8:1, and the same ratio shows up for the relation of helium-neon to all other noble-gas lasers. Gallium arsenide entries are in the ratio of 3:1 to all other semiconductor lasers.[14] Indeed, whenever a new genus of laser was invented, the efforts of scientists rapidly converged around a few species within it. The argon ion laser quickly took the lead among ion lasers, while the carbon dioxide laser dominated molecular gas lasers. The explosion of research around them came about because they were the lasers that worked best. They were therefore the focus of development and of applications research. They were also the lasers around which new companies formed. Clearly, these are the lasers on which we need to concentrate if we want to exhibit the links between technical novelty and the movement of lasers from laboratory to workplace. In addition to the argon ion laser and the carbon dioxide laser, we shall also follow the history of an important newcomer in the crystal group, the neodymium-doped yttrium aluminum garnet (YAG) laser.

The class of ion lasers was introduced by William Earl Bell, one of the founders of Spectra-Physics and the company's manager of experimental research.[15] Bell had been trying to extend the life of helium-neon lasers, and one of the methods he tried involved introducing mercury vapor into the tube. The bright green glow that formed near the cathode suggested to him that mercury itself might lase. "I had a long (three-meter) pyrex glass tube set up in my lab which was used to study scaling effects in helium-neon lasers . . . . I decided to see what it might do with a buffered mercury discharge. [Bell buffered the mercury with helium gas.] . . . Using the standard glow discharge exciter that made the HeNe go, I ran a whole range of gas pressures and electric currents, but nothing happened. I then decided to discharge a high-voltage capacitor charged by a neon sign transformer through the tube at 120 hertz and—Wow!"[16] Bell almost immediately observed a red-orange line different from any known laser line, and later that day a new green line. Comparison of the wavelengths with spectroscopic tables revealed that the lines were not coming from neutral mercury, as Bell and Spectra-Physics theorist Arnold L. Bloom had anticipated, but from ionized mercury.

The laser community was well aware at this point of the importance of lines of shorter wavelength than red, and of the need for green in particular. The U.S. Navy wanted lasers in the blue-green for underwater communications and for the detection of submarines and mines. It had been sponsoring work at RCA, Lear Siegler, and General Telephone and Electronics to develop devices at these frequencies.[17] Blue-green light, which propagates farthest through sea water, would also be useful for mapping the sea bottom for military and commercial purposes. At the same time, research biologists were asking for continuous-wave (CW) lasers at shorter wavelengths because biological systems are more absorbent in that range.[18] These wavelengths, however, were proving difficult to get, except by second-harmonic generation, which entailed a loss of power. As Bloom explained a few years later, a visible line from a neutral atom such as neon is a freak, because by and large the differences between the energy levels at which lasing occurs in neutral atoms are too small to give a radiated wavelength shorter than infrared. In ionized atoms, in contrast, the average lasing transition takes place between energy levels that are more widely separated, producing visible or ultraviolet light.[19] When Bell's publication appeared, therefore, it stimulated wide interest.

At Yale University, William Bennett redirected his research. He had been planning to attack the problem of short-wavelength lasers by looking at high-frequency transitions in neutral argon and neon, and he had built a large, versatile laser for the work. He shifted to ion lasers after learning of the Spectra-Physics result.[20] At Perkin-Elmer, the management asked J. Dane Rigden, then working for the company's English branch, to return to Connecticut to lead a program on mercury ion lasers.[21] At Hughes Research Laboratory, Earl Bell's ion laser caught the attention of William B. Bridges, a research electrical engineer. Bridges's work proved particularly important.

Bridges, who had joined at the Hughes Aircraft Company in June 1961, had been working full-time on gas lasers since mid-1962. It was then that laboratory management, on the heels of the Bell Laboratories discovery of the helium-neon visible line, converted the group to which Bridges belonged into a full-time laser group.[22] Bridges had worked on the development and application of the helium-neon laser. He had also studied the mechanism of operation of a helium-xenon laser invented at Bell Laboratories, and he had built a high-gain pure-xenon laser and studied its application to an optical radar system, under an Air Force contract.[23]

Bell's paper caught Bridges's attention because of the color of the mercury ion lines, because of their relatively high power, and because of the scientific puzzle the work contained, namely the role, if any, that helium played in causing the population inversion. This problem was like the one Bridges had tackled for the xenon-helium laser.[24] He therefore put together a helium-mercury ion laser in his laboratory. As he reported in a *Lasers & Applications* interview,

We built one of these helium-mercury lasers, got it operating in the lab, and then started playing around with it to see what its characteristics were. Is helium essential or not? . . . What's the pumping mechanism? One way to determine that is to take the helium out and substitute neon. . . . That works. . . . After we had made a neon-mercury laser, we decided we would try argon as a buffer gas to see if we could make an argon-mercury laser. Later, we found out that works, too. But this particular day we couldn't make it work. We were using much too much argon. Nor did we have it well adjusted.

So, we pumped out the tube, flushed it, and put helium back in to make sure that our mirrors were still aligned. To our surprise, we found we had a new line going in what was ostensibly a helium-mercury laser. We now had a blue line at 4880 angstroms in addition to the red and green lines from mercury. Well, that was very exciting and totally unexpected.

We measured the wavelength to within a couple of angstroms with a small spectrometer we had. Then it was hurriedly off to the library while the laser was still running. I fumbled around in the library, because I was unfamiliar with the spectroscopy of ionized argon and didn't know where to look. But after about an hour, I concluded that we were looking at ionized argon.

This discovery was made on February 14, 1964.[25]

If there was an element of the accidental in Bridges's discovery of the argon ion laser, there was nothing accidental in the alacrity with which he followed it up. Bridges measured high gains for the argon lines, and that indicated a potential for substantial power. The fact that not only the helium-mercury ion laser, but also the neon-mercury ion laser and then the argon ion laser, had operated meant that the excitation mechanism in Bell's laser had not been the transfer of energy from helium to mercury by virtue of a coincidence of a helium with a mercury resonance, but, probably, direct electron excitation. Resonance coincidences were rare events, but electron excitation was commonplace. Bridges put aside his other projects for the time being; by the end of March, he had made pulsed ion lasers of krypton and xenon as well as argon and had demonstrated a total of 31 new lasing transitions. The gain in the argon 4880 Å line was the highest yet demonstrated for any line in the visible region. In one of his pulsed lasers, the power output was already over 5 watts.[26] Section head Donald C. Forster wrote to notify Air Force contract monitors of this breakthrough that had been made on company research money.[27] Meanwhile, Bennett and his group at Yale and Guy Convert and his coworkers in France independently operated pulsed argon ion lasers (figure 5.1).[28]

The high power of the argon ion laser brought new U.S. groups into the field. One was the Optical Maser Group located in Bell Laboratories' Microwave Tube Department. Eugene I. Gordon, supervisor of the group, had decided not to follow up Bell's helium-mercury ion laser work because he had judged that the mercury ion laser was never going to be anything but pulsed. The Microwave Department was within Bell's development area, where the applicability of the work to AT&T technology was of more direct concern than in the research area. Gordon was directing his group's efforts toward single-frequency continuous-wave gas lasers suitable for optical communications: "I had this conviction that some day . . . optical communication is going to be important—probably with gas lasers—and so I was doing all the things that needed to be done to make it into a practical technology."[29] Gordon learned about the argon ion laser when Bridges gave him a preprint at an IEEE committee meeting they were both attending. He decided that this might be a candidate for a continuous-wave laser. Its power would be in the right ballpark too, 0.1 watt or more output, and it would have the added advantage that most of the available photodetectors for optical receivers were more sensitive to the blue and green portions of the spectrum than to the red portion where the helium-neon laser's visible line was located.[30] Gordon concluded that the key to making a continuous argon ion laser would be a high current density, and this would require a short tube with a small bore. Gordon returned to his New Jersey laboratory and, working with group member Edward F. Labuda, built a quartz laser tube of about 7 inches with a bore of about 1 millimeter. They found it necessary to add a water-cooling jacket to keep the quartz from overheating and a spiral return tube in parallel with the main one to equalize the gas pressures at the cathode and anode. Gordon was able to telephone Bridges by early April, to tell him that he and Labuda had an argon ion laser running CW.[31]

What made this heady stuff was the fact that the continuous-wave powers were also considerable. Bridges, who now joined Gordon and Labuda in work on the continuous argon laser, obtained 150

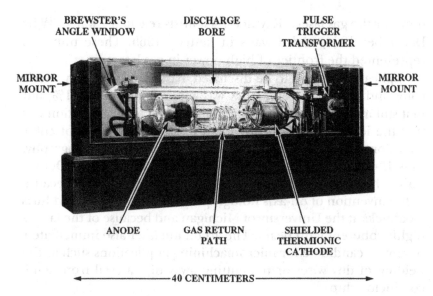

BREWSTER'S
ANGLE WINDOW

DISCHARGE
BORE

PULSE
TRIGGER
TRANSFORMER

MIRROR
MOUNT

MIRROR
MOUNT

ANODE

GAS RETURN
PATH

SHIELDED
THERMIONIC
CATHODE

◀———————— 40 CENTIMETERS ————————▶

Figure 5.1
Pulsed ion argon laser built at the Hughes Research Laboratories for public
display in July 1964. The power output was a few watts peak in 50-microsec-
ond pulses, with an average output of a few milliwatts in several blue and
green wavelengths. (Courtesy Hughes Research Laboratories, photo by Cliff
Olson)

milliwatts from his discharge.[32] Such a high power so early in the
game presaged the possibility of watts and even tens of watts. In
contrast, earlier gas lasers, such as the helium-neon, could produce
about 100 milliwatts.[33]

One of the teams that picked up immediately on the argon ion
laser was Roy A. Paananen and his coworkers at the Raytheon
Research Division. To that time, the Raytheon team had concentrated its gas laser work around the helium-xenon laser because it
allowed high-power single-mode operation. The Raytheon group
was also beginning a small project on the highly corrosive mercury-
sodium vapor, from which it hoped to get visible lines. The argon
laser, however, was both visible and, potentially, powerful. Raytheon
decided to concentrate as much effort here as it could spare from
its contract work. By the end of October 1964, the group had
achieved an output of 4 watts. "This represents about 1000 times
higher power output per unit volume of gas used as compared to
the previously used neutral gas lasers. . . . It is quite clear that we are
witnessing a breakthrough that will open up entirely new applica-

tions for the gas laser." Raytheon scientists reached 7 watts CW by December 1964, and 8 watts in January 1965. These numbers represented the frontier of high power in gas lasers.[34]

Applications began to be discussed, based on the argon laser's color and power. (Its efficiency was abysmal, less than 0.1%, but that still made it the most efficient ion laser.) One suggestion was that the ion lasers might provide a basis for new types of color television and for visual displays. Another suggestion was holography. This 15-year-old technique, invented by British physicist Dennis Gabor, had begun to enjoy a resurgence in 1962 and 1963 because of the invention of off-axis holography by Emmett Leith and Juris Upatnieks at the University of Michigan and because of the laser's highly coherent light beam. The argon ion laser also immediately became a candidate for micromachining applications such as the welding of tiny wires or the cutting away of material from semiconductor chips.[35]

At Hughes Research Laboratories, William Bridges found himself deeply involved almost immediately in applying the argon ion laser to nighttime battlefield reconnaissance. Night reconnaissance was carried out from planes using a television camera and a source of light. Conventional light sources, such as flares or strobe lights, were cumbersome and easily detectable by the enemy. Perkin-Elmer had therefore pioneered a surveillance system that used a camera fitted with a helium-neon laser. The laser's power output was too low, however, to meet the military's airspeed and altitude requirements. When argon lasers promised to remove these restrictions, Perkin-Elmer shifted to them. Hughes, for its part, now decided to enter the area, judging that argon ion lasers had finally made such surveillance systems practical.[36]

Another application explored was retinal surgery. The pulsed ruby photocoagulators then in use had the disadvantage that their red light was only slightly absorbed by blood vessels. Yet diseases of the vasculature were the second largest cause of blindness in the United States. One pioneer in this field was Francis A. L'Esperance, a New York ophthalmologist. He was already using ruby photocoagulators in his practice, and in early 1965 he turned his attention to argon. He got a 10-watt laser from Raytheon and then worked with Eugene Gordon's group at Bell Laboratories and the Special Products Division of American Optical Company (which was marketing ruby photocoagulators) to develop the optical-mechanical systems necessary to deliver the laser's energy to the eye.[37] So while in the case of aerial reconnaissance the argon laser was transforming an impractical laser application into a workable one, in the case of photocoagulation it was transforming an application already on the market into one that was more widely useful.

At least one American company was formed around the argon ion laser. Orlando Research Company was established in the summer of 1966 by three scientists at Martin-Marietta's Orlando Aerospace Division. William McMahan and James Bowen had participated in building an argon ion laser at Martin-Marietta's Electromagnetics Laboratory on a DoD contract. When they couldn't persuade their management to field ion lasers, they recruited John Tracy and moved their talents into a storefront in a shopping strip. There, next door to a barber shop, they began producing argon ion lasers at the rate of one or two a month.[38]

Hughes Aircraft Company also began to market argon ion lasers in 1966. Raytheon placed ion lasers on the market in 1966, Spectra-Physics and RCA introduced models in 1967, and Coherent Radiation began selling them in 1968.[39]

The early commercial argon ion lasers were unreliable and woefully short-lived. It is worth taking a brief look at the problems of this laser, and the development work done to overcome them, for it will show what was required to make laser tubes application-worthy. The argon ion laser was less intrinsically tractable than the helium-neon laser. The discharge was run through a small capillary tube, typically 1–10 millimeters in diameter. Since high currents were needed, first to ionize the argon and then to establish the inverted population, the net effect was an enormous current density: 100–2,000 amperes per square centimeter for pulsed lasers, as compared to densities in the range of 0.05–0.5 ampere per square centimeter for helium-neon. Special cathodes had to be developed to produce the large currents. Once achieved, the extreme current densities threatened to heat the tubes to melting. The high concentrations of very energetic ions also directly attacked and destroyed the tube wall material, while the tube molecules that sputtered into the discharge functioned as contaminants. The physical mechanisms by which the population inversion was created were, at first, unknown. As laser scientists unraveled them, it became clear that the power output varied with the square of the current density. Since higher power was the name of the game, it was essential to increase current densities. This, in turn, exacerbated the tube and cathode problems.[40]

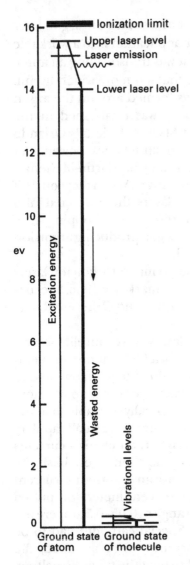

Figure 5.2
The upper and lower energy levels (here measured in electron volts) of an infrared atomic laser are situated far above the ground state. Hence the wasted energy will be a large fraction of the excitation energy. This contrasts with a molecule, in which the lasing transition lies between two different vibrational levels belonging to the ground state.

To solve these problems, a major development effort began into tube structures that could dissipate heat effectively. This was coupled with investigations into tube materials that would be more resistant to thermal damage and to energetic particle bombardment. Tungsten, graphite, and beryllium oxide were among the many materials tried. Solutions became the basis for proprietary designs and could even provide the technical basis for the formation of new companies.[41] In 1970, however, six years after the argon ion laser was discovered, tube lifetimes were still under 1,000 hours, compared to tens of thousands of hours for helium-neon lasers.[42]

The second important laser developed in the mid-1960s was the carbon dioxide (CO_2) laser. It was also discovered in 1964, although the feature that was to electrify the laser community—the combination of high power and high efficiency—was not demonstrated until 1965. The laser's U.S. inventor, C. K. N. Patel, had undergone a conversion experience of sorts when he had attended the International Quantum Electronics Conference in Paris in February 1963. Patel had arrived with hundreds of new laser lines to report on. But at the conference, he recalled in a 1984 interview, "I came to the conclusion that finding just another new line was not the important thing to do in quantum electronics anymore. One had to do something beyond that, namely . . . lasers that had something special about them . . . something other than just yet another line which produced a few milliwatts of power." The reason, Patel explained, was that "quantum electronics by that time was beginning to move away from just lasers to applications of lasers to science." Nonlinear optics was especially prominent at the meeting. "It was very clear that if one were going to use the gas lasers for that kind of scientific stud[y], milliwatts won't get us too far. . . . I came back from Paris with a definite feeling that I had to go to a different kind of system where powers were high. [A] second aspect that was pointed out [was] the question of efficiency."[43] Patel was at that time most interested in nonlinear optics as a source of scientific phenomena, but it is worth remarking that much of the experimental work of others in this area was in fact motivated by technological interests, including interest in finding ways to modulate and demodulate optical beams for communications systems.[44]

Returning to Bell Laboratories' Electronics Research Department, Patel decided that the kind of gas laser he had heretofore researched, which used transitions between different electronic states of neutral atoms, would always be low-powered. His laboratory notebook for spring 1963 contains proposals of various kinds for higher-power lasers. Patel also began to frequent the library, studying molecules. Molecules, unlike atoms, are capable of vibrating and rotating. A molecule in a given electronic energy state can exist in one of various vibrational states, so that each electronic level is subdivided into vibrational levels. Each vibrational level is further subdivided into rotational levels.[45] Comparing this rich level structure with the structure of the levels in atoms, Patel concluded that a laser operating on transitions between different vibrational levels belonging to the ground state in a molecule would be intrinsically more efficient than an atomic laser producing the same wavelengths (figure 5.2). Other data, on the lifetimes of excited vibrational levels in molecules, suggested to him that diatomic molecules would be unfavorable. Patel therefore fixed on carbon dioxide, as one of the simplest and most stable of the triatomic molecules,[46] and on water vapor (H_2O). In fact, Mathias and Parker at the Services Electronics Research Laboratory in England had just gotten diatomic molecules to lase, first the nitrogen molecule in the near infrared, at the extraordinary power level of over 100 watts, and later, roughly contemporaneous with Patel's first experiments on CO_2, the carbon monoxide molecule, in the red, orange, and green parts of the spectrum.[47] But these were pulsed lasers on electronic transitions rather than continuous-wave lasers on vibrational levels. Patel successfully ran a CO_2 laser in an electric discharge tube in late January 1964, on vibrational transitions.[48] Carbon dioxide lasers now became one of a range of topics in lasers and infrared radiation that he pursued.

In about June 1964, however, Patel tapped into another scientific tradition, that of the physical chemists. Reading studies by McGill University scientists on the chemical and physical properties of nitrogen excited in gas discharges, he learned that excited nitrogen molecules can relax by transferring their energy to N_2O, CO_2, or argon.[49] From Patel's perspective, what was interesting in this process was not the relaxation of nitrogen but the energizing of carbon dioxide. In the fall of 1964, Patel showed that he could increase the efficiency of his CO_2 laser tenfold, to 1%, by exciting his molecules by resonant transfer from excited nitrogen molecules.[50]

Through these experiments, Patel's path converged with that of the French physical chemists François Legay and Nicole Legay-

Sommaire. The Legays and their collaborators had been investigating the luminous emissions that carbon dioxide and carbon monoxide emit when they are excited by the transfer of energy from nitrogen.[51] The Legays were also aware, as Patel was not, of a suggestion that the chemist John Polanyi, of the University of Toronto, had published in a chemical journal in 1961, for molecular lasers on vibrational transitions.[52] Building on Patel's CO_2 laser and on Polanyi's suggestion, the Legays suggested independently of Patel the N_2-CO_2 system, and they followed up Patel's experimental results by getting 300 milliwatts from an N_2-CO_2 laser. The extra power was made possible by adding a little oxygen.[53]

Patel, meanwhile, had started working intensively to optimize the geometry and parameters of his laser. More and more, this effort crowded other topics out of the pages of his laboratory notebooks. Before the end of May 1965, he achieved 12 watts CW output at 3% efficiency, using water vapor as an additive, and simultaneous excitation of the CO_2 by both electric discharge and the transfer of energy from excited nitrogen. By the fall of 1965, using helium instead of water as the additive, he had powers of 106 watts CW at 6%, and 183 watts peak pulsed.[54]

A few months after he passed the 100-watt mark, Patel heard informally that the Department of Defense was starting classified research to scale up CO_2 lasers to kilowatt levels. Patel, who was not a citizen yet and had no clearance, realized that he would be unable to follow this research or to muster the kinds of personnel or money resources that the DoD projects would enlist. But while the research field of developing CO_2 lasers to higher powers was getting crowded, the field of applying the laser as a research tool was wide open. Only a handful of research laboratories possessed a CO_2 laser that they could use for research. Since Patel was interested in science rather than development and was convinced that the best way to build a career in science was to initiate new fields, he now turned his attention to the use of CO_2 lasers in nonlinear optics and spectroscopy.[55]

The CO_2 laser now replaced last year's favorite, the argon ion laser, at center stage. In his 1966 review of gas lasers, Arnold Bloom called it "by all odds, the most important development [within the past year]."[56] By then, the CO_2 laser had also been Q-switched, the first gas laser to show this capability, and that bespoke the possibility of peak powers in pulsed CO_2 lasers almost 10^4 times the CW powers. Whereas the argon laser had been the highest-power *gas* laser that had been achieved through early 1965, the CO_2 laser had enough power to rival the ruby and neodymium-doped glass lasers. In Bloom's words, "Until recently, experiments involving high energy density . . . were almost entirely the province of the solid-state laser. . . . The high-power CO_2 laser, however, has changed this picture completely."

The applications that were proposed reflected this development. One was beam weapons. After 1967, the CO_2 laser became the basis for a resurgence of interest in laser antimissile weaponry at ARPA and tactical weaponry at the rest of the Department of Defense.[57] Another application was communications. The CO_2 laser's wavelength, at 10.6 microns, coincided with a window in the atmosphere where absorption was minimal. The properties of high power and low atmospheric absorption also made it a good candidate for radar. In addition, the CO_2 laser's kilowatt powers suggested a way to go beyond the micromachining envisaged for the argon ion laser and carry out full-scale welding and cutting operations; here, however, the CO_2 laser also had drawbacks: A wavelength of 10.6 microns would not be well absorbed by metal surfaces, while the size of the spot to which the beam could be focused—which is a function of wavelength—would be 20 times larger than an argon spot.

Once again, commercial production began almost immediately. The new company of Coherent Radiation, Inc. (today Coherent, Inc.) was one of the first to bring CO_2 lasers to market. Coherent was an offshoot of Spectra-Physics, which had looked at the CO_2 laser but had decided not to market it.[58] In May 1966, three Spectra-Physics employees—physicist James L. Hobart, mechanical engineer Wayne S. Mefferd, and sales manager Eugene L. Watson—left to form a company based on the new laser; they were joined by electronics specialist Robert J. Rorden from Varian and physicist Stephen Jarrett from TRG. Coherent placed its first CO_2 laser on the market in the fall of 1966 and made its first delivery in early 1967.[59]

The third of the 1964 lasers we shall consider, the neodymium-doped yttrium aluminum garnet (Nd:YAG) laser, also came out of Bell Laboratories. It was invented by Joseph E. Geusic in the Solid State Maser Department.[60] Geusic, from the time he had joined H.

E. D. Scovil's maser group in 1958, had had special responsibility for the selection of promising maser materials. In 1962, he had begun a collaboration with Scovil on a review article for the British series *Reports on Progress in Physics*. Geusic and Scovil had used the article to elaborate an analogy Scovil had formulated between an optically pumped maser and a thermodynamic heat pump. The analogy yielded new, somewhat unexpected criteria for selecting the laser materials that would need the smallest possible pumping power.[61] Geusic added to these criteria certain practical requirements from his experience with ruby masers: The material should be hard, for example, and its optical properties should remain as constant as possible under temperature changes. Armed with these standards, Geusic examined experimentally some 30–40 solids. YAG ($Y_3Al_5O_{12}$) was among the group that had the right behavior.[62]

The trouble with YAG, however, was that no crystals that were sufficiently long for lasing were available. Geusic approached Bell's crystal scientists, particularly LeGrand G. Van Uitert of the Crystal Chemistry Research Department. The two first contracted with an outside company for a YAG crystal grown by the Verneuil or flame fusion method. This allowed Geusic and his technician, Horatio M. Marcos, to run a Nd:YAG laser in a pulsed mode. Then Van Uitert, who with his associates had been wrestling with eliminating the lead that was entrapped in crystals made by the Czochralski method of pulling the crystal from a melt, succeeded in making a lead-free YAG crystal 0.25 centimeter in diameter and 3 centimeters long. Geusic and Marcos were able to run that laser CW and to show that Nd:YAG did indeed require comparatively little power—only about one-fifth as much as the laser using the popular crystal calcium tungstate ($CaWO_4$) as the host for neodymium.[63] Geusic and Van Uitert now began a protracted collaboration with the Linde division of Union Carbide to develop larger crystals of better optical quality. The better the crystals they obtained, the more they could do with Nd:YAG. In 1965, Geusic and coworkers demonstrated Q-switching. In 1966, they produced pulses shorter than 10^{-10} second by mode-locking the laser.[64]

The versatility of the Nd:YAG laser made it attractive to entrepreneurs. In 1967, three scientists from TRG founded the firm of Quantronix to exploit this new laser. TRG had been acquired in 1964 by Control Data Corporation, which had subsequently tightened management control, making its new subsidiary into a division and shifting emphasis from research to product sales. The three had left TRG because of management policies, but they had decided to employ their talents in a firm of their own because the Nd:YAG laser provided them with a niche. In the words of Quantronix cofounder and president Richard T. Daly, Nd:YAG "could whistle more tunes and dance more steps" than any laser then on the market.[65]

Other firms also started to market Nd:YAG lasers. One was Holobeam, a company organized in 1967 by venture capitalist Melvin S. Cook around the three product lines of lasers, military electronics, and arc lamps for lighting applications.[66] Others were Korad, Coherent, Control Laser (Orlando Research), and Raytheon.[67] The CW capability, unusual for crystal lasers, at powers that rose steadily through the 1960s to reach above 700 watts by 1970, made the Nd:YAG laser an interesting alternative first to the use of argon ion lasers for micromachining, and then to the CO_2 laser for larger-scale machining.[68] The high repetition rates at which pulsed YAG lasers could be run made it interesting for battlefield applications such as radar and "smart" bombs. Communications formed another potential application.

AT&T itself considered manufacturing Nd:YAG lasers, but management eventually decided that the number of units to be produced would be too small. Hence, the company chose the strategy of sharing information with small companies, such as Quantronix, so as to ensure its access to an outside supply of lasers.[69] The contrast with the invention process in the late nineteenth century is worth noting. Then, independent inventors sold their inventions to large companies or, if their new ideas were too radical, formed their own small companies to exploit them.[70] But the intervening century saw the rise of the industrial research laboratory, whose business was to produce inventions that the parent firm could then decide to exploit or release. And so in the instances of the Nd:YAG laser, the helium-neon laser, and the CO_2 laser, the inventions were made within the research laboratories of large companies and then handed over to small firms for commercial development.

By 1970, YAG was beginning to rival ruby as the most common crystal laser. Good crystals as long as 15 centimeters and with a centimeter thickness had been grown. Krypton pumping lamps, for high power, were supplementing the longer-lived but less efficient tungsten filament lamps. Still, YAG lasers were not yet wholly application-worthy. Lamp life remained inadequate—a few thousand hours with tungsten lamps and about 100 hours with krypton—and the quality of the crystals was not yet adequate.[71]

The mid-1960s lasers discussed so far contrast, in respect of the motives for their creation, with the lasers of 1961–1962. Invention in the earlier period was often driven by a simple desire to get lasing on new lines or in new types of substances. As we move into 1963 and 1964, however, we find that laser researchers could afford to be pickier, and felt impelled to be. William Bridges was attracted by the short wavelength of the mercury ion laser Earl Bell had discovered as well as by the scientific puzzle presented by its pumping mechanism. Patel was looking for gas lasers with high efficiencies and high powers. Geusic wanted a sturdy crystal laser that could be set into oscillation at low pumping powers. The laser community had begun to direct itself toward discovering lasers with specific properties.

This is not to say that the scientists sought to create lasers to serve the needs of specific applications. In this connection, we need to recognize that in places like Bell Laboratories, or Hughes, or IBM, there was a division of labor between those who worked on specific devices such as the laser and those who worked on full systems for functions such as optical communications or radar. Such a division was reflected, for example, in the organization of Bell Laboratories, which was divided into distinct research, component, and systems groups.[72] At smaller companies, at least in the early days, the same person might work on both the component and the system: In the mid-1960s, for example, Spectra-Physics president Robert Rempel worked on both the company's helium-neon lasers and the Geodolite, a surveying instrument that was a major company project.[73] As the 1960s drew to a close, however, small laser companies began increasingly to turn themselves into manufacturers of components, selling their wares to firms called original equipment manufacturers, where the complete systems were assembled, so that a division arose here too. Along with this division of labor went a difference in consciousness. The needs of particular applications were not present to device scientists with the same immediacy that they were to systems engineers.

These needs were instead communicated to the laser device people through a variety of channels. There were informal channels such as conversations, and there were intraorganizational structures such as company seminars. There were also, of course, more formal forums. For example, scientists concerned with applications often made presentations at laser meetings or at meetings covered by the trade press. In October 1963, for example, *Electronics* covered a speech to the American Welding Society in which MIT Professor C. M. Adams jr. explained what welders needed in a laser.[74] He called for more pulses per second for pulsed lasers, a feature that translates into a requirement for greater efficiency,[75] and for the invention of something that did not yet exist: a high-power continuous laser. As another example, in August 1964, Captain Martin S. Litwin of the Army Medical Research & Development Command alerted attendees at the Boston Laser Conference to the fact that researchers in biology needed continuous lasers at all wavelengths, but particularly at wavelengths shorter than 5900 Å.[76] The Department of Defense and NASA arranged special symposia, some classified and some unclassified, to lay their own needs before the laser industry.[77] It scarcely needs saying that systems requirements became ever clearer during the decade, as more and more systems studies and experiments were carried out.

Out of all this, a consensus emerged among laser device scientists as to the kinds of properties that were desirable. Many of these properties were needed for more than one application. High powers were needed for both laser weapons and the industrial processing of materials. Room-temperature operation of semiconductor lasers would make them more useful for both communications and computers. High efficiency was important for most industrial and military uses. Such properties ascended to a meta-applications level, where they became virtues per se; everyone making new lasers was aware of them and hoping to achieve them.[78]

How well does my claim that laser-making became more purposeful as the 1960s advanced fit the facts when we move beyond the three cases of the argon ion, CO_2, and Nd:YAG lasers? It would seem, indeed, that two other important mid-1960s lasers, the chemically excited laser and the dye laser, contradict it. For although the dye laser ultimately proved to have the welcome property of tunability, the researcher who first operated it in the United States, Peter P. Sorokin, was not searching for a tunable laser and was not the one to show how to coax this behavior out of it. Similarly, the first chemical lasers made in the United States were constructed by researchers whose principal interest lay in investigating reaction kinetics rather than in building compact high-power lasers. If we examine the context of these two discoveries, though, we find that a substantial number of scientists were, at the time of the dye laser, deliberately trying to invent tunable lasers or to doctor existing lasers to make them tunable. We also find that a number of scientists were experimenting with chemical reactions as a way of making laser systems more compact. Thus, the cases of the chemical and dye lasers do not negate the generalization that laser scientists were becoming more purposive. Rather, they show that the process was too rich to be completely constrained within any single generalization.

In the development of the chemical laser, chemists as a disciplinary group played an important role. Chemists, in fact, were involved with lasers in several ways in the early and mid-1960s. They were interested in them as a light source for spectroscopic investigations and as a potential means for initiating industrial processes for the chemical industry, and they had a professional concern in the CO_2 laser and its relatives, whose active media were composed of larger molecules. The latter was something of a jurisdictional matter. The chemists were willing to allot atoms and diatomic molecules to the physicists, but triatomic and bigger molecules they saw as part of their own turf.[79]

Even before Maiman announced his ruby laser, the Canadian chemist John Polanyi had formulated the idea of a chemically pumped laser. He pointed to the fact that the products of a chemical reaction may be born into excited states.[80] Consider the reaction $XY + Z = X + YZ^*$. The asterisk signifies that the molecule YZ is in one of its possible excited states rather than its ground state. Consider also two particular excited states of YZ, YZ_1^* and YZ_2^*, that are suitable as lower and upper laser states, respectively. If the chemical reaction produces more YZ_2^* than YZ_1^*, it produces an inverted population, and under suitable circumstances lasing can take place. The pump itself produces the laser medium, or, expressed from the opposite angle, the laser is self-pumping.

George C. Pimentel, in the Department of Chemistry at the University of California, Berkeley, knew Polanyi and had read his 1961 article. He and his group had built a ruby laser in 1962, because Pimentel thought that in one way or another lasers were bound to become an important source of light for spectroscopic measurements.

In early 1964, Pimentel initiated research into chemical lasers. This work was an outgrowth of his long-standing interest in the properties of transient species formed during chemical reactions. As part of that research program, in the early 1960s Pimentel had assigned graduate student Kenneth C. Herr the task of building a spectrometer that could rapidly chart the near-infrared radiation from excited molecules. Herr's spectrometer, complete in early 1964, could record in 200 microseconds all the radiation emitted after a reaction in the spectral region between 1 and 15 microns. It was two orders of magnitude faster than any preceding apparatus.[81] The success of Herr's spectrometer induced Pimentel to look at the problem of making a laser, and he put another graduate student, Jerome V. V. Kasper, to work on one suggested by Polanyi, where the reaction is $Cl_2 + H \rightarrow HCl^* + Cl$.[82]

The radiation Kasper got from the hydrogen-chlorine reaction was inexplicably inconsistent, and the experiment was further clouded by an extraneous phenomenon—the appearance of shock waves in the apparatus. In spring 1964, Pimentel had Kasper put hydrogen chloride aside and turn to the photodissociation of compounds of the form CX_3Z, and where X is hydrogen or a halide and Z is some other atom or group. They expected that the CX_3 would emerge from the dissociation vibrationally excited, and Kasper, working with CH_3I and CF_3I, adjusted the spectrometer to look for CX_3's characteristic infrared emissions. What he got was a series of perplexing signals that he finally elucidated by showing that they were emerging, not from the CX_3 group but from the iodine, which had been born into an electronically excited state. This itself was a result of chemical interest. In the 1930s, D. Porret and C. F. Goodeve had predicted that this photodissociation would produce I* in the electronic state denoted $^2P_{1/2}$, but it had not previously been possible to test their forecast. Kasper's result was also technologically interesting, because the gain was so large that the system would lase even when Kasper substituted an nonreflecting surface (an index card!) for one of the mirrors.[83]

It is pertinent to note that while Pimentel and his group were pursuing the chemical laser as an offshoot of their success with the rapid-scan spectrometer, scientists at other laboratories were working with chemical reactions as a way to solve the concrete problem of weight and compactness. The high-power optically pumped lasers then available, and especially the ruby laser, had the disadvantage that they came fettered to massive power supplies. Raytheon's 1962 catalog, for example, offered ruby lasers whose head, containing the ruby crystal, flash lamp tubes, and Fabry-Perot cavity, weighed about 15 pounds, while the cabinet containing the capacitor banks and circuits that provided power to the flash tubes weighed 300 or 400 pounds.[84] Hence researchers were investigating whether optical pumping might be effected by using chemical reactions that could produce intense luminescence instead of flash lamps fired by electrical power supplies. North American Aviation had a $95,000 contract from the Office of Naval Research to investigate flares made of mixtures of metals and oxides. Scientists there talked of reducing overall weight by as much as a factor of 25. Scientists at the U.S. Army Picatinny Arsenal in New Jersey and the Stanford Research Institute in Menlo Park, California, were also exploring methods of producing light of enormous brightness from chemical reactions.[85]

Chemical lasers also held possibilities as space-based beam weapons. Two chemicals might be carried aloft in separated compartments in a rocket and combined above the atmosphere to give the reaction. No external power source would be required; mere combination would provide the pumping energy. For this reason, as well as for their intrinsic interest, three scientists attending an Institute for Defense Analyses (IDA) study group in Woods Hole, Massachusetts, in July 1963 decided to organize a conference on chemical lasers; the three were Kurt E. Shuler, of IDA and the National Bureau of Standards, Keith Brueckner, at that time vice president of IDA, and William Bennett.[86] Their idea was to bring together experts on chemical kinetics and on lasers so as to stimulate chemical laser research. Even though no chemical laser had been made to work when the conference was being organized, enough theoretical and experimental work had been done to schedule more than 20 papers.

The Conference on Chemical Lasers was held at the University of California at San Diego in September 1964. It was funded by the Air Force Systems Command and the Office of Naval Research. While none of the scheduled talks at the La Jolla conference disclosed an operating chemical laser, Kasper attended and was able to report informally that he had just gotten his iodine photodissociation laser running in the week preceding the meeting.[87] Soon after, Kasper returned to the hydrogen chloride reaction and demonstrated lasing there.[88]

The chemical laser was one of those proposed prior to the operation of the first working lasers.[89] No chemical laser was operated in this first period, however, and the quest continued into the mid-1960s. At this point, a push from the side of applications was added. Chemical lasers looked interesting as a candidate for beam weapons. The Air Force and Navy sought to stimulate chemical laser research through their funding of the La Jolla conference. All the while, the chemists had an agenda of their own. For while chemical knowledge offered a new route to lasers, lasers also offered a path to new chemical knowledge. Research on chemical lasers thus had two markets for its wares, one within the discipline of chemistry and one within technology. The disciplinary market was to remain the principal one until 1969, when continuous-wave chemical lasers were invented. At that point, the center of gravity began to shift from using the chemically excited laser to study chemistry to using it in technological, and especially military, applications.[90]

The dye laser, first operated in late 1965, proved to be one of the most important of the tunable lasers.[91] Tunability was a property to which substantial attention was being paid at this time. It was sought both through research on existing lasers, exploring the extent to which they might be made tunable, and through proposals for, and experiments with, new devices that would have the property. Yet the dye laser's inventors, Peter P. Sorokin and John R. Lankard at IBM, were not looking for a tunable laser when they found it, and it took several years before tunability emerged as preeminent among its properties.[92]

Tuning means continuously varying the wavelength of a (pulsed or continuous-wave) laser through some given range.[93] Tunability was an obvious desideratum, quite apart from any particular application, because it was a property of radiofrequency and microwave sources. One constant underlying animus of the whole laser undertaking was the desire to extend all the properties of radiofrequency sources up the electromagnetic spectrum toward higher frequencies.

There were also pressures for tunability from particular applications. Spectroscopy, both an area of scientific research and a potent industrial tool for monitoring chemical composition, was one source of pressure. Microwave spectroscopists used the tunability of microwave sources in their investigations of molecules. At the time of their initial work, in 1957 and 1958, Townes and Schawlow already foresaw that optical masers might be used in the same way for spectroscopy in the visible and infrared.[94] Although ingenious methods of using the high intensity and monochromaticity of single-frequency lasers were developed during the 1960s,[95] the lack of tunability was sorely felt by spectroscopists. Theodor W. Hansch, who was to become one of the major figures in laser spectroscopy in the 1970s, remembered of his student days in the mid-1960s, "We had these marvelous ideal light sources, much more monochromatic than anything that people had known before, and it was also clear that there were new ways of studying spectra of atoms with much better resolution, but you couldn't tune these lasers to wavelengths that were interesting. I remember that much of my time I felt a sense of frustration."[96]

As the new field of nonlinear optics developed, investigators in it also were increasingly held up by the lack of tunable sources. Nonlinear effects needed powerful sources, in the range of 10–100 megawatts. Only ruby, at 0.7 micron, and neodymium-doped glass, at 1.06 microns, sufficed. Other frequencies could be generated from these, at somewhat lower powers, by second-harmonic generation or related methods. But writing in 1965, the Soviet scientists S. Akhmanov and R. V. Khokhlov explained, "such frequency converters for laser energy are far from capable of solving the problem . . . [which] can be regarded as solved only when the frequencies of the coherent oscillations will be made continuously variable just as, for example, in the microwave band of the electromagnetic spectrum."[97]

Attempts to design tunable, coherent optical sources were already in the literature by 1962. In that year, Norman M. Kroll at the University of California, San Diego, and Akhmanov and Khokhlov in the Soviet Union independently proposed the tunable "variable-parameter" or "parametric" oscillator.[98] Parametric oscillators were already well known at lower frequencies. To create an optical version, Kroll and the Soviet scientists envisioned use of a fixed-frequency laser, such as the ruby laser, to illuminate a crystal with markedly nonlinear properties. Because of the crystal's nonlinearities, the energy of the pumping beam could be transferred into two other electromagnetic vibrations within the crystal whose frequencies depended on the pumping frequency and the crystal's index of refraction.[99] Tuning could then be achieved by varying the index of refraction in any of a number of ways. Kroll suggested varying the angle of the crystal with respect to the direction of the oscillations. Other suggestions were soon added: varying the temperature, or pressure, or applying electric fields. In 1965, Joseph Giordmaine and Robert C. Miller at Bell Laboratories took advantage of a new crystal that one of their colleagues had just grown, with exceptionally strong nonlinear coefficients, to demonstrate the first optical parametric oscillator.[100]

How wide a net was being cast for tunability can be illustrated by a number of examples drawn from U.S. laboratories. In 1963, Dietrich Meyerhofer and Rubin Braunstein at the RCA Laboratories in Princeton examined tunability in the newly invented gallium arsenide laser, and they succeeded in varying the output wavelength over a range of 20 Å by applying a mechanical force perpendicular to the pn junction.[101] In 1965, a trio of engineers from Stanford University analyzed a proposal for an infrared Raman laser to show that such a laser might be capable of tuning over tens of angstroms.[102] In 1966, Leo F. Johnson and two colleagues at Bell Laboratories broke through to a new order of tunability. Their laser, specially constructed of transition-metal ions (that is, nickel, cobalt, etc.) in crystals of titanium dioxide and perovskite fluorides, could be tuned by almost 2000 Å in five discontinuous segments. Continuous tuning over the longest segment was 250 Å.[103]

In contrast, the first dye lasers came not out of this research tradition, but from the investigation of Q-switching. By early 1964, the methods for Q-switching had been extended from Kerr cells and rotating mirrors to the use of uranium glasses or of thin films of organic dyes deposited on glass plates. These methods work by the "saturation" or "bleaching" of an absorption. When light is sent through an absorber, energy is removed from it because molecules in lower energy states are excited into higher ones. If sufficient light is applied, however, the supply of lower-state molecules is used up and the absorber, normally opaque, at characteristic frequencies becomes transparent (or "bleached" or "saturated"). It now transmits unimpeded the light it previously absorbed. At IBM, Peter Sorokin and his group showed in 1964 that a simpler and more convenient bleachable absorber for Q-switching could be made from a group of dyes called metal phthalocyanines, dissolved in liquid organic solvents.[104]

Once that experiment was completed, Sorokin later wrote, "We were to benefit further from having on hand generous amounts of the . . . phthalocyanine compounds prepared for the bleachable dye absorber experiment. . . . Their singular spectra led to suggestions for other experiments. . . . One idea was to try to produce stimulated (resonance) Raman scattering. . . . Another thought was to attempt to obtain stimulated emission from the dyes, using a ruby laser as a fast flash lamp. . . . We initially opted to try the first experiment. When the spectrum of light emitted from the ruby-laser-irradiated cell was examined, it was apparent that we had instead succeeded with the *second* experiment: We had observed stimulated emission from the dye."[105] The independent discovery of dye lasers in Germany stemmed from the same research tradition and was even more of an accident, according to an account by Fritz P. Schaefer: "At that time the author, unaware of Sorokin and Lankard's work, was studying in his laboratory, then at the University of Marburg, the saturation characteristics of saturable dyes of the cyanine series. . . . When Volze, then a student, tried to extend these measurements to higher concentrations, he obtained signals about one thousand times stronger than expected. . . . Very soon, . . . it became clear that this was laser action."[106]

The path between this accidental discovery of the lasing of dyes useful for saturable absorbers and the appreciation of dye lasers as tunable sources par excellence stretched over the rest of the decade. In 1966, Sorokin and Schaefer both used Q-switched ruby lasers as pumping sources to excite the dyes (figure 5.3).[107] The laser photons they got from the dyes were necessarily lower in energy, and therefore longer in wavelength, than the roughly 6940 Å of the ruby. The dye beams, moreover, had a broad spectral output, in the range of 100 Å, because each vibrational level has a rich substructure of rotational levels. The peak of this broad band could be varied by hundreds of angstroms by changing any one of a number of parameters, such as the concentration of the dye within the solvent or the length of the tube within which the solution was contained.[108] To get high-intensity pumps at smaller wavelengths, it was possible to use second harmonic generation of ruby or neodymium beams, getting pumping wavelengths at 3470 and 5300 Å, respectively. This made it possible to use dyes with higher-wavelength output. Sorokin, for example, reported, "We got the second harmonic of our ruby going, and we just began pouring

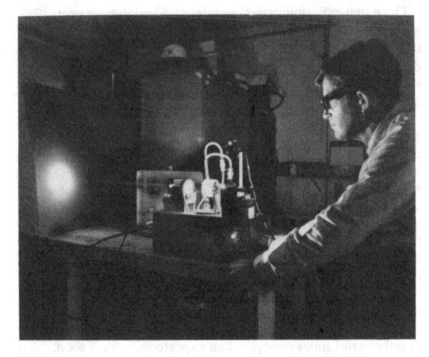

Figure 5.3
Peter Sorokin firing a flashlamp-pumped dye laser, 1968. (Courtesy IBM Thomas J. Watson Research Center)

various fluorescent dyes into a cell between two mirrors and seeing which ones lased. I remember one afternoon we just went down the aisle here [in the laboratory] asking our colleagues, 'What color do you want?'"[109] A more important answer, however, was Sorokin's invention of a fast, high-intensity flash lamp.[110] Its use meant that by early 1967 it had become possible to get gross stepwise tuning over the entire visible spectrum.

In mid-1967, another decisive advance was published by two scientists at Korad, Bernard H. Soffer and Bill B. McFarland (figure 5.4). They substituted a reflecting grating for one of the laser mirrors and were able both to tune the center frequency over more than 400 Å and to narrow the spectral width by a factor of 100; for one dye they used, this was a decrease in width from 60 Å to 0.6.[111] At the same time, they retained 70% of the energy output, now funneled into a narrow line. Narrowing was rapidly improved. Then, in 1970, continuous-wave dye lasers were added to the stock of pulsed ones.[112] By the early 1970s, the dye laser, though still unable to compete with older fixed-frequency lasers for uses that

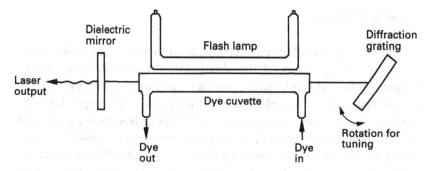

Figure 5.4
Soffer and McFarland replaced one mirror with a diffraction grating. The grating functioned as a mirror reflecting a narrow band of frequencies around a frequency peak that was altered by rotating the grating. (Adapted from B. B. Snavely, "Flashlamp-Excited Organic Dye Lasers," *Proc. IEEE* 57 (1969), 1387)

did not demand tunability, had emerged as the major tunable laser for the visible region.[113] Schaefer, with a rhetorical flourish, called it "the fulfillment of an experimenter's pipe dream . . . : To have a laser that was easily tunable over a wide range of frequencies."[114]

While the dye laser was in this manner sliding into its special niche in laserdom, the frontal attack on tunable lasers continued with vigor. One example is the work C. K. N. Patel undertook to make the spin-flip Raman laser. Patel and two colleagues had previously demonstrated that the effect—in which an electron in a semiconductor crystal absorbs some energy from the photon of a laser beam to flip its spin, with the residual energy departing as a photon of lesser frequency—is a strong one.[115] The effect was also tunable, because the amount of energy the electron took up changed with the strength of an externally applied magnetic field. In late 1967, Patel put aside other research in order to try using the spin-flip Raman effect to make a high-intensity tunable laser for spectroscopy in the infrared. He succeeded in this effort in 1970.[116]

The serendipity of the discoverers of the dye laser thus exists against a backdrop of a deliberate search for the property they found by accident. There is, of course, nothing unusual in the fact that there were surprises and accidents in the story. That is part of the fabric of science. There was also an element of playfulness, something much more permitted to scientists in the 1960s than in the more tightly controlled 1970s. Sorokin's dye laser work fits under this head and shows how much such playfulness redounded to the profit of lasers.

Laser Applications

Laser applications claimed increasing attention as the 1960s drew to a close. Organizers of the third, 1971, Conference on Laser Engineering and Applications, for example, noted that the proportion of systems studies submitted to them was increasing relative to studies of the laser component, and that this systems work was increasingly experimental, as opposed to mere proposals.[117] The index to *Electrical & Electronics Abstracts* for 1969–1972 shows how wide was the variety of applications studies. The more than 1,000 abstracts listed under "laser beams: applications" ranged from laser canes for the blind and laser dentistry to fire detection, from spectroscopy to the alignment of linear accelerators, from measuring the heights of ocean waves to measuring the properties of semiconductors. By far the greatest number of papers, however, dealt with three topics: laser communications, the machining of materials, and the measurement of lengths of all sorts, from the dimensions of workpieces in industrial processes to the tiny strains that constitute the diurnal earth tides.[118] Behind these frontrunners came another group of subjects represented by significant numbers of papers: the measurement of the velocities of fluids; the monitoring of constituents of the atmosphere and especially atmospheric pollutants; information storage; the tracking of planes, missiles, and satellites; biomedical applications; and the study of properties of plasmas, which are gases so highly energized that a large fraction of their particles have been dissociated into electrons and ions.[119]

The frontrunners—communications, machining, and length measurement—were each driven by a combination of technical opportunity and demand, but the way in which these factors combined was different for each of them. For optical communications, at which we look first, the driver was the urgent need for greater transmission capacity for electronic communications. The telecommunications industry, the military, NASA, and all of their contractors felt this need. Telephone use was soaring. The relaying of television signals now took up a large segment of transmission capacity. Electronic transmission of facsimiles and of messages between computers was spreading. The command and control requirements of the military were growing, while the move into space had opened a wholly new set of communication needs. Optical communication by coherent laser light seemed at first an ideal solution. As early as 1963 and 1964, however, substantial technical problems became evident. A laser communications revolution that had once appeared imminent now was assessed to be some years away. Discouraging comments became common. Yet researchers pushed on, even, at times, with technical systems that were clearly going to be clumsy and expensive.

The initial attraction of the laser for communications engineers was its short wavelength. The amount of information a coherent electromagnetic wave can carry, under ideal conditions, increases with its frequency. Optical waves have frequencies about 10^9 times larger than radio waves and about 10^5 times larger than microwaves. It appeared that coherent optical waves might be able to carry more telephone calls, television programs, facsimiles, and data than all the radiofrequency and microwave equipment then installed. A second perceived advantage was that the laser beam can be collimated into a tight pencil of rays that barely spreads as it travels. This meant that laser messages could be sent over long distances with relatively low transmitter power, since a far greater fraction of the emitted beam's energy would arrive at the receiver. Tightly collimated beams would also have advantages for secret communications, since less of the information would spread out to eavesdroppers. A third advantage was the narrow spectral width of laser light, which meant that spectral-band filters could be used to block out much of the light of incoherent background sources of "noise" such as the sun.[120]

The earliest laser system put into use for communications involved short-range, line-of-sight laser links. It was becoming increasingly clear that the atmosphere had deleterious effects on laser light traveling through it: Incoherent fluctuations were impressed on the beam that could not be compensated for, no matter how much ingenuity went into the design of the receiver.[121] Nevertheless, for distances of a few miles, atmospheric effects were not disabling. As early as 1963, experimental systems for use on missile test ranges or battlefields were reported in the trade press and engineering journals. The first entrants were from electronics companies with laser experience, including General Electric, Hughes, RCA, and Sylvania GTE; later systems were introduced by manufacturers, such as Mark Systems, Inc., whose traditional products were nonlaser military optics. These systems combined in one package transmitter and receiver and were generally made with gallium arsenide lasers. They were far more compact than the microwave systems they replaced; they were, essentially, augmented binoculars with weights in the range of 3–9 pounds. They could not, of course, be used when bad weather limited visibility. Moreover, the gallium arsenide lasers deteriorated with time. In the late 1960s, scientists were beginning to elucidate the character of this degradation process but were not yet able to overcome it. Nevertheless, by 1969, laser communications links were already being tested on the battlefields of Vietnam.[122]

The use of lasers for space communications progressed far more slowly than short-range land links. The advent of spacecraft in the late 1950s had created a demand for new kinds of telecommunications. Communication channels were needed to command spacecraft from earth and to relay data to earth from military reconnaissance satellites, weather satellites, and planetary probes. Communications engineers set out to meet these demands with microwave equipment. They faced problems, however. To go from microwave links spaced 30 miles apart, which was the common distance for terrestrial microwave systems, to space-to-ground links of about 1,000 miles for low-lying satellites, 25,000 miles for geosynchronous satellites, or millions of miles for deep-space probes, seemed to require microwave tubes of very high power. But high-power tubes and their power supplies were heavy loads for spacecraft, and the highest-power tubes were also short-lived. Microwave equipment also had the drawback that the rates at which data could be sent back was low for some purposes. For interplanetary probes, for example, the rates were so low that ingenious means of storing the data on the spacecraft and relaying it back to earth at sloweddown rates were needed.[123] In this situation, a number of organizations began to advocate laser space links. For interplanetary probes under ideal conditions, laser light could be expected to transmit information at high rates, perhaps 1 million bits per second as opposed to the 10 bits per second that could be achieved with microwaves. One would expect optical antennas and optical components to be smaller and lighter than microwave apparatus. Optical equipment would also be less susceptible to interference from high-altitude nuclear explosions. Finally, the high degree of collimation of laser beams meant that less energy would be wasted, so that transmitters of lower power would be feasible.

NASA mounted a major laser space communications project in the mid-1960s, spearheaded by an enthusiastic team at the Marshall

Space Flight Center at Huntsville, Alabama. Perkin-Elmer and Chrysler Corporation were among Huntsville's major contractors. The Marshall Center group used a continuous-wave helium-neon laser as their transmitting source.[124] This project, and other smaller ones conducted by aerospace firms, ran into a series of technical problems. The most serious were those associated with directing the narrow laser beam to make contact with the receiver on the spacecraft and keeping it there as the craft moved ("acquisition and tracking").[125] This problem was especially severe for the helium-neon laser that NASA used, because of its low power. A second major problem was the atmospheric degradation of light. What was tolerable for a short-range land link of a few miles became formidable for the hundreds of miles of atmosphere that space-ground links encompassed. This drawback led some researchers to propose that lasers be used only between deep-space vehicles and earth satellites and that microwave frequencies then carry the information from the satellites to ground stations.

Still another problem was that of developing components for use with the laser. The modulator was the worst bottleneck. It is the component that impresses the information that the laser radiation is to carry. Improved modulators began to be developed out of new materials in the middle 1960s. But these still suffered from the problem that the amount of information they could impart to a light wave was small. "Under ideal conditions" in the phrase "the laser has enormous information-carrying capacity under ideal conditions" means that the other components in the communications system must match the laser's information capacity.[126]

In the late 1960s, NASA's Goddard Space Flight Center and the Air Force each began working on communications systems using the new higher-power lasers, notably the Nd:YAG laser and the carbon dioxide laser.[127] But these lasers were still short-lived and bulky. By the end of the decade, no laser space system had yet been flown, and microwave and millimeter waves were still considered the best carriers by space communications engineers.[128]

The last major area of communications research during the 1960s was that of terrestrial optical links, including short intracity or even intrabuilding links on the one hand and long-distance, high-capacity intercity links on the other. Communications companies started programs in many major industrial countries. In the United States, the largest program was at Bell Laboratories.[129]

Systems work centered in the Guided Wave Research Laboratory, directed by Stewart E. Miller, and was part of the Research and Communications Sciences Division, where Kompfner was an associate executive director. Throughout the 1960s, Kompfner continued to play the role of prodder and evaluator of Bell's laser communications research.[130]

Bell scientists made many studies of the degradation suffered by light sent through the atmosphere, in clear weather and in rain, fog, and snow. Early on, they formed the conviction that, for long-haul systems at least, the light would have to be sent through some form of closed tube. Bell Laboratories, and Miller's laboratory in particular, were heavily involved at the time in a program to design a transmission system that used millimeter waves propagated through waveguides.[131] It is probable that their technical ideas on transmitting optical waves were influenced by this parallel research program.

Shielded transmission posed the problem that optical elements, such as glass lenses, would be needed along the path to confine the light within the tube. Still more lenses would be needed to direct the light along curved paths, and the tubes would have to be curved if for no other reason than to follow the earth's curvature. There would inevitably be a large loss of energy at the lens surfaces, however, and by the mid-1960s the laboratory staff shifted its attention to a gas lens invented by staff scientist Dwight W. Berreman. Here, slowly flowing gas was kept denser in the center of the tube and rarer at the periphery by an application of heat at the tube's circumference.[132] Gas lenses eliminated surface energy loss but were like glass in that the transmission tubes had to be built to stringent requirements and would be expensive. The justification for pursuing such unpromising technologies was the hope of using light to get enormous information-carrying capacities.[133] AT&T was spurred, not only by the generally mounting need for communication channels, but also by an invention of its own, the Picturephone®. Bell officials had high expectations for public adoption of the Picturephone, which permitted customers to see as well as hear their telephone callers. But a Picturephone connection required 250 times the channel capacity of a simple telephone call.[134]

Engineers working on terrestrial communications also faced the problem of designing optical components, and above all modulators, that could complement the light beam's information-carrying capacity. On the face of it, semiconductor lasers appeared to offer a major advance, since they could be modulated both easily (by superimposing alternating currents on the direct current used for pumping) and rapidly. However, semiconductor lasers were not yet reliable, stable, or powerful enough. Nor were they yet capable of operating continuously at room temperature. Their output was in the infrared, whereas at least for tests of experimental systems, it was more convenient to use visible wavelengths. AT&T communications engineers took the view that while semiconductor lasers were bound to be of major importance at some point in the future, for the present, optical systems had to be built around the more developed, and more reliable, gas laser.[135]

In the late 1960s, communications engineers turned some of their attention from continuous-wave lasers to pulsed solid-state lasers. The technique of mode-locking solid-state lasers had by then resulted in the production of regular trains of high-power pulses, each lasting for times on the order of picoseconds (10^{-12} second) and separated by times on the order of a nanosecond (10^{-9} second). Scientists began to think about combining these ultrashort pulses with pulse-code modulation (PCM), in which information is encoded in binary form, as patterns of signals signifying "yes" or "no" (or, alternately put, as patterns of 0s and 1s). Impressing information on a train of laser pulses would then become a matter of transmitting some pulses and suppressing others. For picosecond pulses, whose duration was a thousandth of the interval between two successive pulses, this opened the possibility of interleaving many different messages ("time-multiplexing"), and this, in turn, would finally provide a way of exploiting the laser's potential for high information rates.[136] But this project remained, as the 1970s dawned, just another item on the research agenda.

Until 1970, terrestrial links appeared to many specialists outside the Bell system as the least promising of all the communications applications of lasers. The difficulty of transmission through the atmosphere and the clumsiness of alternative solutions such as closed tubes filled with glass lenses or flowing gas formed a principal objection. Thus, Raytheon's Clarence F. Luck jr. wrote in 1964 that lasers "do not seem to offer any obvious advantages" for terrestrial communications, whereas "the space communication application . . . does offer considerable promise." Bernard Cooper of ITT Federal Laboratories in 1966 judged that "the perturbing effects of the earth's atmosphere . . . raise a serious obstacle to . . . earthbound communication systems," as opposed to space and interplanetary communications. The following year, James Vollmer of RCA reckoned lasers to be helpful "for distances less than about 15 km [i.e., short atmospheric links] and for those greater than 80 million km [i.e., space links]."[137] Indeed, the failure of lasers to bring on a rapid revolution in terrestrial communications was one factor behind the erosion of the euphoria that had marked the early 1960s.

In 1970, two technical breakthroughs occurred that dramatically changed the situation. First, building on results of Charles K. Kao and George A. Hockham of Great Britain's Standard Telecommunication Laboratories, Robert D. Maurer and his group at the Corning Glass Works fabricated a glass fiber that could transmit light with a loss of only 20 decibels per kilometer.[138] (The number of decibels measures the ratio of the power fed into the fiber to the power that emerges.) At that time, Bell Laboratories' best loss figure for glass fibers was over 400 decibels/kilometer, and Bell's emphasis was still on hollow waveguides with lenses.

The Corning achievement transformed the rules. Transmission through low-loss optical fibers had the potential advantages of low cost and a transmission medium less bulky than copper cables. Optical communications, from being a technology restricted to ultra-high-capacity links, now appeared as one that could be competitive in the near future with coaxial cables and other communications media.[139]

The second breakthrough was in the area of semiconductor lasers. A new technique for manufacturing semiconductor pn junctions, called liquid-phase epitaxy, led to the production of heterojunction lasers, formed of alternating layers of gallium arsenide and aluminum gallium arsenide. First, single heterostructure lasers were made, reducing the current density needed to reach the threshold for lasing from more than 100,000 amperes per square centimeter for the earliest homojunction lasers to 8,000 amperes/centimeter^2. Then, at the end of the 1960s, double heterojunction structures were made, reducing the threshold current density to the range of 1,000–3,000 amperes/centimeter^2 and thereby for the first time permitting continuous operation at room temperature (figure 5.5).[140] The wavelength and size of the new semiconductor lasers were well matched to the

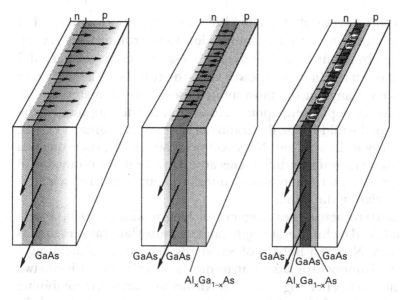

Figure 5.5
The double heterostructure laser (right) restricts both light and charge carriers to a smaller volume than either the single heterostructure laser (center) or the homostructure laser (left). This allows the lasing to take place at a lower charge density.

properties of fibers. The year 1970 thus ushered in an entirely new era for terrestrial optical communications.[141]

The other two frontrunning applications, laser interferometry for the measurement of displacements and laser materials processing, are worth pairing for comparison. Both were envisioned at an early stage. Townes, in his July 1958 proposal to the Air Force Office of Scientific Research for funding for the Columbia University laser work, picked out the "marked improvement in interferometry and measurements of length by interferometric techniques" as one benefit to be anticipated from the research he was proposing. George Smith at Hughes Research Laboratories in July 1960 placed micromachining among eight applications for the laser Maiman had just invented that were worth exploring.[142] Both applications were studied assiduously during the 1960s, but the state of the laser technology upon which each drew was different, and the contrast between them is therefore useful for understanding how the maturation of an application depended upon the maturation of the laser component.

The interferometric measurement of length used principally the visible helium-neon laser invented by White and Rigden in 1962.

This was a relatively tractable laser, and the improvements that were indispensable for its application to metrology were accomplished by about 1968. Materials processing, on the contrary, did not have in the first high-power laser, the ruby laser, a satisfactory source for any but the most limited uses. Instead, materials processing was dependent upon the invention and development of the lasers of the mid-1960s, the argon ion laser and, to a greater extent, the carbon dioxide and Nd:YAG lasers. Materials processing was also more technically difficult because it involved the interaction of light and matter, whereas laser interferometry was purely a matter of regulating the light.

Interferometric measurement of length was, in 1960, a well-established technique at weights and measures laboratories such as the U.S. National Bureau of Standards. It involved generating a beam of monochromatic light, using a "splitter" to divide it into two beams, then reflecting those beams from two mirrors, recombining them, and analyzing the interference pattern they produced. By moving one of the mirrors along the length to be measured and counting the cycles of the interference patterns (they would go through a full cycle every time the mirrors moved another wavelength apart), one could use the known wavelength to calculate the length.[143] The magnitude of lengths that could be measured depended on the degree of monochromaticity of the light source. Before the laser, the most monochromatic source available—a green line of the mercury isotope of atomic weight 198, from which all but a narrow range of frequencies had been filtered—produced measurable interference patterns over a length on the order of a meter.

After the laser was invented, the National Bureau of Standards began research into laser interferometry. In 1962, Bureau scientists demonstrated fringes over a 200-meter path, using a helium-neon laser. Several technical problems had to be addressed, however, in order to apply laser interferometry to length measurement. First, the laser had to be operated in a single mode, with a single frequency. Second, the wavelength corresponding to that frequency had to be accurately measured. Third, the laser had to be frequency-stabilized.[144] The need for frequency stabilization can be appreciated by recalling that the frequency of a laser depends on the length of the Fabry-Perot cavity. The cavity length, however, can be altered by changes in the ambient temperature and pressure, by airborne vibrations, and by mechanical vibrations transmitted from the building to the laser through its supports.[145] The Bureau's earliest interferometric laser used a rigid quartz spacer to hold the two Fabry-Perot mirrors a fixed distance apart. Later apparatus used a feedback control based upon the so-called Lamb dip, which is a dip in the curve relating the output power of a gas laser to the frequency at which it is operating. This dip was so named because it had fallen out of the equations that Willis E. Lamb jr. formulated when, in 1961 and 1962, he elaborated a model for the behavior of gas lasers. Subsequently, the physical cause of the dip had been elucidated, and its existence had been experimentally verified.[146] The dip provided a marker at a particular frequency, and deviations from this frequency could be converted into electrical signals that could be used to readjust the positions of the cavity mirrors.[147] By 1967, the National Bureau of Standards and laboratories abroad were using laser interferometry routinely in some of their length measurements.[148]

Interferometric measurement of lengths was not common in industry in 1960. It was used in the optics industry in the manufacture of ruled gratings and to some extent in the calibration of measuring rulers. But the short coherence length of the light sources limited the magnitude of the lengths that could be measured, while their intensity was so attenuated by the use of filters that it was impossible to use interferometers in well-lighted areas.[149] The laser made the difference. Several companies now initiated research on the industrial applications of laser interferometry. Frequency stabilization was, again, one of the technical problems that was tackled. The lifetime of lasers became interesting, since it would have been purposeless to install a piece of equipment in which the laser component would have had to be replaced every 100 hours or so. Speed of measurement was another factor that took on far more significance in an industrial setting.

In the United States, Airborne Instrument Laboratories had an industrial laser interferometer system, using a Spectra-Physics helium-neon laser, for sale by the end of 1964. By 1966, another New York company (OPTOmechanisms, Inc.) and a midwest company (DoAll Company) were also offering systems. Airframe manufacturers, who were finding it necessary to make ever longer parts to ever narrower tolerances, and machine-tool companies, which needed to verify that their numerically controlled machines could position themselves accurately at the coordinates being specified by the tapes that controlled them, were early customers. In 1966, Boeing Aircraft Company had two laser interferometers in use to calibrate the motions of its machine tools and to check the gauges that measured the dimensions of the workpieces. The Cincinnati Milling Machine Company used a single laser interferometer to check its automatic tape control machines. In addition to their superior accuracy, the laser interferometers had the advantage of a speed that far outpaced conventional length-measuring techniques of the late 1960s. At this point, however, laser interferometers were 30–40 times as expensive as other kinds of measurement apparatus, and they were therefore not used for routine measurements.[150]

Through the next few years, research groups worked to develop interferometers that could be mounted directly onto the machine tools and used to control and measure their motions. Oak Ridge engineers developed a set of compact instruments, suitable for mounting on lathes and milling machines, which they costed out at $12,000, of which $6,000 was for the laser. Perkin-Elmer marketed an industrial interferometer for calibrating measuring scales, while it looked forward to the ultimate use of its instrument as a "machine control loop feedback element." The Bendix Corporation was developing a variety of interferometers, including noncontact probes for measuring the inside and outside diameters of bearings and rings.[151] Hewlett-Packard entered the market in 1970, through a Scientific Instrument Division it established in May 1970.[152] By the end of the 1960s, we can summarize the status of laser interferometric length measurement as follows: The stabilized helium-neon laser that was its core component was a mature device. A number of companies were selling systems for routine use in production situations. These systems were not yet being bought in quantity, however, but in the ones and twos. The total market for laser interferometers was in the range of $10–15 million yearly.[153]

Unlike laser interferometry, which was taken up by standards laboratories and was also a useful experimental tool in seismology and other branches of science,[154] laser materials processing was a wholly industrial application. It did not enjoy the luxury of being first explored in scientific laboratories, where cost considerations were not of prime importance and where the physical environment was lenient and performance standards less exacting. Corresponding to its purely industrial status, work on laser materials processing fell, in the United States, within the purview of laser vendors and research production engineers in major companies.

Many laser companies initiated work early on. Korad was one of the earliest vendors to pursue materials processing. It received a major contract to develop a laser welder from Linde, a subsidiary of Union Carbide, the company from which Korad had received its capitalization. Linde was already a major supplier in the welder market, and the possibility of developing a laser welder was one factor in Union Carbide's decision to invest in Korad.[155] Westinghouse set up a Scientific Technologies Division in 1964 and included in the division a small group devoted to laser systems. This group worked on laser welders and drillers for use within the company and for sale to outside firms.[156] The commercial ruby laser unit at Hughes introduced a laser microwelder at a meeting of the Institute of Electrical and Electronic Engineers in March 1964.[157] Lear-Siegler's Trion Instruments Division worked on laser welders, and General Electric, TRG, and Raytheon all flirted with them.[158]

Among production engineering groups, AT&T's Western Electric Engineering Research Center in Princeton, New Jersey, was one of the earliest and most prolific contributors. Honeywell's Corporate Research Center was another. Scientists in these laboratories spent much of their time investigating the way in which laser beams interact with materials. Indeed, because of this interaction, laser materials processing was intrinsically more difficult, technologically, than laser interferometry. In interferometry, the adoption of lasers was a matter of plugging a new light source into a proven system and then improving the source to meet the needs of the system. In materials processing, in contrast, the laser was to be substituted for very different methods of working materials, such as mechanical drills, electron beams (then a 10-year-old technology), or electrical resistance welding. Laser beams affected the workpiece very differently than these older methods. Their effect was also very complicated. Research revealed that it depended on the properties of the material being worked—its thermal characteristics, surface condition, chemistry, and metallurgy—as well as on the properties of the laser—its wavelength, power density, and the size and shape of the focused spot. As the interaction proceeded and the material of the workpiece changed phase, melting and/or vaporizing, the character of the beam-matter interaction also changed. Moreover, the ambient atmosphere and the alterations induced in it by the laser also affected the results. The nature of the process led to requirements on the intensity, spot size, and duration of the laser pulses and also on the shaping of the pulse, that is, on the programming of its intensity as a function of time so as to conform to the changing response of the workpiece.[159]

Welding, although it was one of the earliest machining applications investigated, proved one of the most difficult to implement. The lasers with sufficient output power to be useful, ruby and neodymium-doped glass, had low repetition rates. Hughes's 1964 microwelder, for example, pulsed at only 9–12 shots per minute. Welding a seam by a sequence of laser pulses this slow took far more time than welding by conventional methods. Laser welding was therefore prohibitively expensive for seams.[160] To do welding, it is necessary to control the power density at the workpiece carefully so that the area being worked is kept at a temperature high enough that it melts but low enough that it does not vaporize. This kind of control was difficult to achieve.

The first laser metalworking technique that did go on a production line was drilling rather than welding. At the end of December 1965, Western Electric introduced the laser drilling of holes through diamond dies into its Buffalo, New York, plant. The diamond dies were used to manufacture fine wires for electrical connections. Each wire had to be drawn through a succession of 14–18 dies, so that the plant used several thousand dies in its operations. Piercing the die, or resizing it, was done with metal drills or diamond dust. It could take up to 24 hours, and breakage was high.[161]

The die-drilling system was created at the Western Electric Engineering Research Center. A major component of the work was safety testing. The effects of laser beams on eyes and skin, or of plumes of vaporized materials on overall worker health, were not yet established. "There was so much that was unknown about it," Sidney S. Charschan, a research leader at the center, recalled, "that you were not going to be allowed to eat in an area where there was a laser for fear of radiation contamination." Western Electric decided on overkill. The Engineering Research Center designed a bright, ventilated workroom with an enclosed workstation that the operator guided from the outside by closed-circuit television. Another closed-circuit television monitored the operator.[162] It took Raytheon's Special Microwave Devices Operation, which built the laser and system to Western Electric specifications, 7 months of close collaboration with the Engineering Center to complete the machinery. The final laser system performed the preliminary rough drilling of the dies in a matter of minutes and proved a strikingly cost-effective process.[163] Yet even this clearly superior application spread only slowly. According to the trade magazine *Laser Focus*, the second U.S. die-drilling laser system only went on-line, at a Pennsylvania plant of the Fabricated Metals Division of the H. K. Porter Company, in March 1968.[164] Laser drilling with ruby lasers still had technical problems. The massive power supplies that fired the flash lamps were too large to be conveniently accommodated on factory floors. Flash lamps were short-lived and had to be replaced frequently.

Another area closely related to drilling was the trimming of microcircuit components, such as resistors and capacitances, to tolerance by the evaporation of bits of material. Here there were jobs that a laser could do that no other machining technique could accomplish. Trimming resistors encapsulated in glass was one; until the laser came along, these resistors were manufactured, tested, and simply discarded if they did not meet specifications.

To this point, laser systems were viewed as having a specialized and restricted, though highly useful, role to play in machining. Lasers could process hard-to-work materials such as diamond and some ceramics. They could machine through glass and other transparent materials. They were being introduced into microelectronics. Their high cost and clumsy structures, however, meant that the benefit to be reaped had to be large if their introduction was to be profitable.[165]

Throughout the mid-1960s, scientists involved with materials processing eagerly awaited each new high-power laser. The argon ion laser, the first of these, created a moment of excitement. Because its repetition rate was high, it was expected to weld at about 100 centimeters/minute, in contrast to the rate of roughly 1 centimeter/minute for the pulsed ruby laser. Its short wavelengths meant that light could be concentrated into a smaller spot than the red and infrared light of ruby or neodymium-doped glass lasers, with attendant higher power densities. Its respectable power output, in the 1–10 watt range, was well suited to micromachining.

Processing scientists, however, soon shifted their interest from argon to the high-energy carbon dioxide laser, which came along a year later. Carbon dioxide lasers had efficiencies on the order of

10%, as against 0.1% for argon, and this was an important consideration in industrial lasers. The power output was much higher, and there seemed to be no end in sight to how high it could rise, with longer lasers giving better figures almost daily. Its long wavelength, 10.6 microns, was a disadvantage. For one thing, it implied a large spot size. For another, metals are strongly reflecting at that wavelength. But the power of carbon dioxide lasers nevertheless portended a breakthrough, from the micromachining of materials less than a tenth of a centimeter thick to the macromachining of slabs 2–3 centimeters thick. By the end of 1967, Nd:YAG lasers also began to claim attention for microprocessing. The Nd:YAG laser was lower in power and efficiency than carbon dioxide lasers, but it was far more compact and had a wavelength one-tenth as long.[166]

In contrast to the case of laser interferometry, much of the activity in laser materials processing, in the half decade from 1967 to 1972, was in basic research. Processes had to be developed that took advantage of the strengths, and allowed for the weaknesses, of the new lasers. Thus, for example, laser manufacturers began to use the beam of the carbon dioxide laser, which was so poorly absorbed by metals that gold film, for example, reflected 95% of it, to cut nonmetals, such as cloth, according to programmed patterns.[167] Another important development was laser-assisted oxygen jet cutting of metals. (In this process, the laser heats the metal to ignition and the oxygen jet combines with metal in a rapid exothermic reaction, while it simultaneously sweeps the oxides away from the workpiece. The width of the cut depends not upon the diameter of the jet, but upon the much smaller diameter of the laser spot, so that a savings of material is possible.) By 1970, this process, which was pioneered in Britain in 1967, was being seriously studied by U.S. aircraft companies for use with expensive airframe metals such as titanium.[168]

It is also instructive to compare materials processing with battlefield applications such as rangefinders, target designators, and illuminators. Today, these are both leading sectors of the market, with dollar sales of the laser components of materials-processing systems roughly twice that of military systems.[169] Military tactical systems, however, preceded materials-processing systems into routine use by 5–10 years. At least part of the timing difference was a result of the intensity with which R&D was pursued and the nature of the markets.

Each of the three armed services had equipment needs shaped by its distinct mission. The Navy was interested in underwater communications, as well as ship-based radar and optical communications. The Air Force wanted to increase the accuracy of its bombs and needed radar to locate the enemy missiles and planes that threatened its bombers. The Army wanted radar for tanks and for hand-held weapons. Each, therefore, actively supported laser research in its focal area. As the Vietnam War expanded, more and more money was allocated for laser research, and, in addition, the services set up special procedures to expedite the work. The Air Force for example, set up Project 1559, a "quick reaction" fund to take low-cost weapons to the point of practicality within 6–12 months. Laser-based guidance systems, which gave tenfold increases, and better, in the accuracy with which bombs were delivered, were developed within this framework.[170]

In contrast, materials processing received little military money until the early 1970s, when laser machining began to prove itself for aircraft materials such as titanium.[171] In the 1960s, materials-processing research was mainly financed by industry. Some of the groups that pursued it worked out imaginative financing schemes. Westinghouse's Scientific Technologies Division, for example, arranged to share the cost of its research between its parent company and the outside firms for which the division was exploring particular applications.[172] The very existence of this kind of creative solution, however, brings home the fact that materials-processing research subsisted through the decade on a modest budget.

As to market structures, it can be said in a preliminary way that one factor that influenced the rate of dissemination of laser materials processing was the technical sophistication, or lack of sophistication, of its customers. The superior technical know-how of the semiconductor industry was probably a major reason why semiconductor resistance trimming was one of the first markets that developed. The fact that AT&T had the Western Electric Engineering Research Center facility was crucial to AT&T's putting on-line the first machining application. Another pioneer in laser machining technology, General Motors, had the advantage of its Warren, Michigan, Technical Center and its Saginaw Steering Gear Division Manufacturing Research Group.[173] Not all firms in the market had resources of this type to draw on. In contrast, it is probably fair to say that military procurement agencies, both within the Department of Defense and in armies abroad, were more uniformly knowledgeable.

Furthermore, the civilian laser industry from which the materials-processing firms were buying machines was more unstable, around the turn of the decade, than were the defense contractors, and the materials-processing machines—lacking the rigid requirements laid down for defense procurement—were more unreliable. One commentator, reflecting on the early 1970s, wrote: "Many of the early high power 'industrial' lasers earned reputations for high downtime, varying power output, and complicated operation or maintenance procedures.... [A] credibility gap between laser vendors and users developed ... [that] was further aggravated by the continuously changing cast of players in the laser world. Companies came and went; laser divisions were bought and sold; new machine designs were frequently announced; and technical battles waged among vendors left potential users confused and frustrated."[174]

When Hughes Aircraft delivered an order of 300 rangefinders for the M60AI-E2 battle tank at the end of the decade, the event was hailed in the trade press as the first "mass market" for laser systems.[175] Hyperbole no doubt, but understandable when we recall sales of industrial lasers were running at a rate of 1 or 2 per buyer.

All the time that laser systems for civilian and military technologies were being investigated, lasers were also being applied to science. As early as 1961, two Bell Laboratories scientists, S. P. S. Porto and D. L. Wood, showed that a ruby laser whose crystal was cooled, to maintain it at constant temperature, could be used as a light source in Raman spectroscopy,[176] where tunability is not a requirement but very bright and monochromatic light is important. Shortly thereafter, Townes and Javan, now both at the Massachusetts Institute of Technology, and two others teamed up to use the laser in a test of special relativity theory. Their experiment focused on the theory's assumption that light has the same velocity in every system in which it is measured, whether or not that system is in motion with respect to the "fixed" stars. Townes had previously collaborated on similar experiments using the maser.[177] The laser version rested on the circumstance that the frequency of a laser beam depends on the velocity at which light travels within the laser cavity[178] and on the fact that the frequency of the helium-neon laser could be determined to such a high degree of precision (1 part in 10^{13}) that extremely small changes in the velocity of light would lead to measurable changes in laser frequency.

The scientific use of lasers spread from laser physics to other fields.[179] Laser measurement of particle velocity provides one example. In 1963, Herman Z. Cummins and some coworkers in the laser group at the Columbia Radiation Laboratory started to study macromolecules in dilute solutions by scattering the light of a helium-neon laser from the solutions and looking for frequency changes induced by the macromolecules. This problem had been suggested to them by a Columbia University chemistry professor. They found a spreading of frequencies that they could attribute, in part, to the motion of the molecules.[180] This and follow-up investigations led biologists to adapt the method to the study of the motion of bacteria; by the early 1970s, sophisticated computer methods had been devised for determining the distribution of velocities of swimming bacteria and their response to environmental variations.[181]

New methods of shaping laser beams and new lasers were promptly put to use. The development of picosecond lasers made a tremendous impact. Many molecular processes, such as the relaxation of excited states or the transfer of energy between molecular species, are played out in intervals of tens or hundreds of picoseconds. Such processes had previously been inaccessible to investigation. Beautifully ingenious experiments began to be devised. In one design, a laser pulse lasting a few picoseconds would be used to excite a substance that would be left to decay under natural conditions while a second pulse, split off from the first at the laser's exit window and made to travel a different path of variable length, would be used to operate a shutter (or "optical gate"). In this way, the fluorescence that accompanied the decay of the substance could be sampled, for times as short as 8 picoseconds, at any desired point in time, and the variation in the intensity of fluorescence with time, a variation that contained keys to the substance's behavior, could be charted.[182]

Tunable lasers were taken up by scientists who were both laser physicists and spectroscopists (or who became both) and were developed to a point at which they were sufficiently monochromatic to be useful in high-resolution spectroscopy. Arthur Schawlow had moved from Bell Laboratories to Stanford University in 1961. In the late 1960s, his laboratory became a center for the improvement and use of tunable dye lasers.[183] Theo W. Hansch, who joined Schawlow's group in 1970 as a NATO fellow, succeeded by 1971 in making tunable dye lasers with outputs as monochromatic as 0.004 Å. In the early 1970s, he, Schawlow, and the Stanford group were doing laser spectroscopy of hydrogen atoms that was accurate enough to cast new light on issues at the foundation of quantum mechanics and quantum electrodynamics.[184]

Lasers in a Changing Environment

At the end of the 1960s, the laser industry's economic prospects began to change. There were at least two aspects to this. First, the market for lasers seemed to be starting a long-term shift from being predominantly military to being largely civilian. Military spending for laser hardware and laser R&D contracts continued to rise, but at a slower rate than before, and the Department of Defense's share as a laser customer fell from 63.4% in 1969 to 58% in 1970 and 55% in 1971.[185] Second, one of the important classes of customers for lasers, university researchers, was entering a period of (relative) impoverishment. The Mansfield amendment, passed in 1969, prohibited defense agencies from supporting basic research not related to their missions. The National Science Foundation, which was supposed to take up the slack, instead found its budget cut. Fred P. Burns, then president of Apollo Laser, later recalled that "in 1970 the lights went out in the science market."[186]

Meanwhile, scientists in industry found their scope for fundamental laser research curtailed, as companies redirected R&D toward projects that held promise for short-term payoffs.[187] Anthony J. DeMaria, a leading researcher at United Technologies Research Laboratory, described the change eloquently:

In the 1960s there was a tremendous national technical renaissance. . . . It was a glorious decade. . . . I was young and motivated by doing research in my field of interest. . . . When you're making important contributions to your field of science, you are treated like a star by the research community. You go to conferences and other researchers recognize you. You get invited to give presentations all over the world, etc. You're riding high when you're productive . . . . [The] young spirit keeps pointing out to you all the contributions one could make to your field. . . . Our group was operating more like the philosophy of a university department. [In contrast, in the 1970s] government money began to stay constant. The in-house money stayed constant. Inflation went up. The overhead went up because of new capital equipment, so we were beginning to be strapped for funding. So the reality of the 1970s came in. . . . I had to think more like a manager with responsibility for people's jobs and to serve the needs of the corporation rather than science. I had to dedicate less and less of my time to science and more and more on how to find funds by being relevant to the divisions.[188]

In tandem with these economic changes came opposition to the Vietnam war, spreading widely in 1968 and after. The impact was strongest in the universities, where laser scientists, like other scientists, were forced to confront their colleagues and their students over the issue of war research. Alexander J. Glass, who had been on the staffs of the Institute for Defense Analyses and the Naval Research Laboratory as a laser scientist before he moved to Wayne State University in Detroit in 1968 to head the Electrical Engineering Department, remembered a vivid incident. He had organized a special symposium on the military's influence on science. As the symposium got under way, it was suddenly transformed into a kangaroo court in which he was one of the accused. "Somebody said: 'Sit down, you're not in charge here.'" Glass reported, "We . . . were put on trial for war crimes. . . . It was really very interesting. . . . One was forced to explain the rationale. 'Why did you do it? Why did you make those choices?' . . . It was impossible to go through that era without really giving some thought and scrutiny to what you were doing."[189]

As a consequence of these economic and ideological changes, laser manufacturers began to adopt new strategies in 1969 and 1970. They moved away from the development of military products and toward civilian products, away from single lasers sold to laboratories for research and toward complete systems for industrial use.[190] Even a company as closely tied to the Department of Defense as Hughes Aircraft joined in, with its manager of press information, William A. Herrman, declaring that he expected civilian laser business to exceed military business in 3–5 years for everyone in the industry, including his own company.[191]

The case of AVCO Everett Research Laboratories (AERL) provides some data for this change in orientation. AERL was set up by AVCO Manufacturing Corporation, a conglomerate with some major military divisions, in 1955, during the heyday of the formation of centralized laboratories. Its mission was basic research beyond the limits of normal company research. Although it was part of AVCO Corporation, it was financed chiefly out of the contracts it secured and the commercial sales it made.[192] Under Arthur R. Kantrowitz, its founding director, it devoted itself to serving national security needs, to scientific excellence, and to operating at a profit. AERL became heavily engaged in military electronics and missile and space technology.[193]

It was AERL that proposed the gas dynamic laser to the Department of Defense in the mid-1960s. By then, research on ruby and glass lasers, which had been the mainstay of the DoD's high-energy laser program, was being abandoned. Patel's carbon dioxide laser was under study, but it could be scaled up to high powers only by dint of prodigious increases in length. (In 1967, the Army Missile Command Laboratories at Redstone Arsenal ran a carbon dioxide laser at the record value of 2.3 kilowatts by using a discharge tube 178 feet long.) The gas dynamic laser that AERL advocated also used a mixture of carbon dioxide, nitrogen, and water vapor, but it obviated the need for long discharge tubes by using a different method of obtaining a population inversion (figure 5.6). The basic idea was to take a hot mixture of the working gases and cool it extremely rapidly. The initial hot mixture would be in a state of equilibrium, with the lower energy states having larger populations

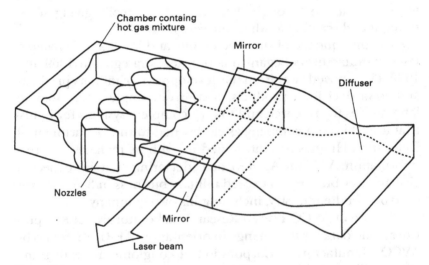

Figure 5.6
Schematic diagram of a gas dynamic laser.

than the higher energy states. The cooled gas would eventually attain equilibrium at its new, lower, temperature. But if it were cooled sufficiently fast, it would be out of equilibrium for a time because different excited states have different relaxation times. In particular, by adjusting the proportions of the gases and the parameters of the system, an excited level of carbon dioxide that might serve as an upper laser level could be "frozen in," that is, made to relax at a very slow rate, while the excited level that was its lower laser level could be depopulated rapidly.[194]

One of the fastest ways to lower the temperature of a hot gas is to expand it through supersonic nozzles. Temperature changes as large as 10^6 centigrade degrees per second can be achieved by this method. Here the laser art unites with the science of aerodynamics, a branch of physics in which Kantrowitz and the AERL staff excelled. In the gas dynamic laser that AERL successfully demonstrated in 1966, the gases remained in a state of population inversion for about a meter downstream of the nozzles. Laser mirrors placed with their normals perpendicular to the line of flow created a resonant cavity. The high gas densities and high rate of flow permitted very large powers; in addition, the flow process itself removed depleted material and waste heat.[195]

The gas dynamic laser was, after some hesitation, taken up with enthusiasm by ARPA and the armed services. AERL and United Aircraft became contractors in a supersecret development program.[196] Nevertheless, toward the end of the 1960s, AERL began to place new emphasis on civilian technologies. Its most important initiative was in medical electronics, where it developed a cardiac assist device based upon its special expertise in fluid mechanics. It also began to explore civilian laser technologies, among them laser pollution monitoring, photochemistry, and materials processing. Kantrowitz recalled, "We saw the DoD research budget going down, or at least not growing, and we, together with I guess the whole scientific community, got kind of tired of defense work. . . . [It was] the spirit of the late sixties, the Vietnam war. . . . We really searched hard for nonmilitary projects, while keeping up the military work."[197] AERL's most striking foray into civilian laser technology was a program on laser separation of uranium isotopes, started in late 1969.

Laser photochemical separation of isotopes was not a new idea. Jean Robieux, who led a laser group at the Centre de Marcoussis of the French Compagnie Générale D'Electricité, and Jean-Michel

Auclair had filed a patent in 1963 for the laser separation of uranium isotopes. They proposed to start with uranium hexafluoride and make use of the fact that some of the spectral lines of the two isotopic forms $U^{235}F_6$ and $U^{238}F_6$ that lay in the infrared were displaced from each other by rather large amounts. They proposed a two-step ionization process. First, an infrared laser would be used to excite $U^{235}F_6$ selectively, and then a second, conventional ultraviolet light source would be applied to ionize the excited molecule. The ionized $U^{235}F_6$ molecules could then be separated out from the $U^{238}F_6$ by virtue of their charge.[198] In the United States, the earliest attempt to use lasers to separate isotopes was reported in 1967 by Arthur Schawlow, Henry Warren Moos, and Stanford graduate student William B. Tiffany. Since at that time lasers could only be tuned through small wavelength ranges, the three were forced to pick a laser line and then search for an element whose spectral lines lay near it, rather than first picking an element with a favorable pair of isotopically displaced spectral lines. They worked with a ruby laser, which they tuned over 9 Å by varying the temperature, and with molecular bromine vapor.[199]

As tunable lasers were developed in the mid-1960s, and as these were refined to give increasingly monochromatic beams, the scientific community reacted with a number of independent proposals for isotope separation. In Israel, Isaiah Nebenzahl presented a secret report to his country's Atomic Energy Commission in early 1969 describing a two-step process using tunable dye lasers and atomic uranium in the form of a vapor. In the United States, Stephen E. Harris, a professor of electrical engineering at Stanford, sent the Atomic Energy Commission and the National Science Foundation a proposal in 1970 to use the tunable optical parametric oscillator to separate isotopes of elements such as bromine and carbon. In Russia, scientist V. S. Letokhov in 1970 took out two patents for the separation of uranium isotopes, one using atomic uranium and the other uranium compounds.[200]

AERL started its own program when staff scientist Richard H. Levy learned from a consultant that French and Israeli researchers had been thinking about the problem. Levy recognized that such a technology would fit well with AERL capabilities. One of the laboratory's research groups was just then developing a tunable dye laser pumped by a nitrogen laser, which AVCO was later to market to research laboratories as the Dial-a-Line laser, and which promised to give wavelengths over the whole visible spectrum.[201] Thus, for a separation process based on uranium vapor, for which visible wavelengths would be needed, AERL would have the head start of having equipment at hand, at least in the initial experimental stages. In addition, commercial prospects appeared excellent. In 1970, the Federal Power Commission projected that U.S. electrical capacity would quadruple by 1990 and that the proportion of that capacity contributed by nuclear reactors would rise from 1.4% to 49.3%. Foreign nuclear capacity was expected to rise even more quickly: The AEC estimated that it would increase a hundredfold by 2000. Ninety-nine percent of the world's need for enriched uranium was being supplied by the U.S. government through its gaseous diffusion plants. Increasing demand, however, was even then placing a strain on the government plants, and the Nixon administration and members of the congressional Joint Committee on Atomic Energy had begun to explore the idea of opening the field to private industry.[202]

The AERL leadership financed its uranium separation project through a long-term arrangement with Jersey Nuclear (later Exxon Nuclear), a wholly owned affiliate of Standard Oil of New Jersey.[203] Jersey Nuclear had already established a study program with General Electric to determine the feasibility of a joint venture in gas centrifuge isotope separation. The laser isotope project would be a second string in Jersey Nuclear's bow, a technology that might be phased in after centrifuge plants were operating.

AERL's pursuit of laser isotope separation unites all the motivations that I claim drove the strategy-makers as they chose among laser applications. First, AERL—and here this means primarily laboratory director Kantrowitz—was responding to changing markets. The defense and government markets were contracting, whereas the demand for nuclear fuel promised to skyrocket. Intimately connected with the market outlook was the fact that R&D funding was available from the oil companies. Second, Kantrowitz and his scientists were reacting to a societal mood. They shared with many university scientists in the late 1960s a sense of unease with military research. Nevertheless, without a third motivation, technological advances in lasers, AERL's uranium isotope separation process would have been unthinkable. It was the advent of the dye laser, and its development into a tunable laser with a narrow frequency width, that provided the indispensable basis for the work.

Laser fusion offers a second example of how economic and social context affected the development of laser technologies reactor technology. Just as a fission reactor uses the same nuclear process as a uranium bomb, so a fusion reactor, were one to be built, would obtain power from the same process—the fusion of deuterium and tritium—that powers hydrogen bombs.[204] For fusion to occur, the deuterium and tritium fuel must be heated to temperatures on the order of 100 million degrees centigrade, where it exists in the form of a gas, or "plasma," of molecules dissociated into their constituent nuclei and electrons. Two types of fusion reactors are presently being considered. In magnetic confinement fusion, the plasma is confined within a reactor vessel by magnetic fields, thereby allowing the deuterium and tritium nuclei time to collide and fuse into helium, with a resultant release of energy.[205] In laser fusion (or "inertial confinement" fusion), a micropellet of deuterium and tritium is irradiated by laser (or particle) beams and converted into a microplasma. Under the right conditions, enough fusion reactions would occur, before the unconfined particles flew apart, to provide the heat energy for economical generations of electricity. Magnetic confinement fusion power research has been pursued in the United States since 1951.[206] Until the late 1960s, however, laser fusion research had only been carried out in the context of weapons work.

This laser fusion weapons research had been done at the Atomic Energy Commission's Lawrence Livermore Laboratory. It had been started on the initiative of Ray E. Kidder, a mathematical physicist working on hydrogen weapons. In the fall of 1961, after Fred McClung and Robert Hellwarth of Hughes Research Laboratories had Q-switched ruby lasers and obtained pulses with peak powers several orders of magnitude larger than those of ordinary ruby lasers, Kidder sat down and made some calculations. He computed that 100,000 joules of laser energy, focused on a few micrograms of deuterium and tritium within an interval of 10^{-7} second, would lead the fuel to ignite, simulating the explosion of a hydrogen bomb. This would for the first time allow weapons physicists to study thermonuclear explosions and their effects under controlled conditions in the laboratory. The giant pulses of McClung and Hellwarth had a duration of about 10^{-7} second and energies of perhaps a few hundredths of a joule.[207]

Kidder took a colleague who was also a friend of Maiman's and went down the coast to visit Hellwarth, McClung, and Maiman in

April 1962. His purpose was to learn if there was any reason of principle that would prevent his getting the high laser energies and short pulses that he had calculated he needed. The conversations disclosed none. One month later, John S. Foster jr., director of Livermore, put Kidder in charge of an exploratory program.[208]

Kidder and his team looked at several types of lasers in the 1960s. They eventually concentrated on neodymium-doped glass and built systems composed of laser oscillators and amplifiers employing an innovative design in which a set of neodymium-doped glass slabs took the place of the usual solid rod.[209] The program ran on a budget of about $1 million a year in the latter 1960s and was funded by the AEC's Division of Military Application.

While Livermore pursued a classified program of laser fusion for weapons simulation, a Soviet laser fusion group under Nikolai G. Basov was publishing its results. In 1968, Basov and his coworkers announced that they had succeeded in creating thermonuclear fusion reactions. They had used a solid lithium deuteride target and a neodymium-doped glass laser whose beam was amplified through five stages to give energies in the range of 10–30 joules per pulse. The evidence that fusion reactions occurred was the appearance of neutrons. The numbers of neutrons were small, however (about 10 per pulse), and their interpretation as products of thermonuclear reactions was not secure. Then, in 1969, scientists at the French Atomic Energy Commission's Lemeil Weapons Research Center announced that they had obtained 1,000 neutrons per pulse from the irradiation of a few cubic millimeters of solid deuterium.[210] The managers of the magnetic confinement fusion program at the U.S. Atomic Energy Commission considered adding a program on laser fusion energy, but they rejected the idea. They estimated that a practical device would require a laser at least 100 times more powerful than that needed for mere weapons simulation, and hence capable of delivering 10^7 to 10^8 joules per pulse as against Kidder's 10^5. Energies that high were too far from the state of the art, which then stood, at best, in the hundreds of joules.[211]

In 1969, however, the American public was also expressing growing concern over environmental pollution. One target of this concern was fission reactors, which were attacked for the radioactive wastes they produced and for their thermal pollution of adjacent bodies of water. Environmentalists were also opposing the

breeder reactors that the AEC, Congress, and the Nixon administration offered as the next generation of atomic generators. Yet
supplies of fossil fuel were palpably limited. In this situation, fusion
energy, hitherto an obscure program within the Atomic Energy
Commission, began to move into the limelight. Its advocates were
making plausible claims that it would have far less deleterious
environmental effects than fission energy and pointing out that the
supply of deuterium for fusion fuel was for all practical purposes
inexhaustible.[212] The growing social attractiveness of fusion energy
combined with the Soviet and then the French laser results to draw
U.S. scientists into the pursuit of laser fusion as a power source.

In January 1970, John Nuckolls and Lowell Wood, two Livermore
scientists from outside Kidder's group, presented the Atomic
Energy Commission with a scenario for laser fusion energy. According to their computer calculations, when a micropellet is irradiated
with laser light, the initial photons create a corona of plasma into
which the subsequent light energy is deposited. The corona trans-
fers energy to the outer layer of the pellet, which can blow off with
such large outward velocities that the rest of the outer layer, by
conservation of momentum, is driven inwards. If the compression
is symmetric and if the pellet remains cool as it implodes, then
densities in the center of the microsphere will rise a thousandfold.
At these densities, the center of the pellet should ignite,[213] and the
burn that started there would spread through the entire pellet.[214]
Nuckolls, Wood, and their collaborators estimated that "central
heating" lowered the energy requirements to 10^5–10^6 joules, and
that "breakeven," the point at which the amount of thermonuclear
energy produced just equaled the amount of laser energy hitting
the pellet, might be as low as 1,000 joules.

There were difficulties. Nuckolls and Wood assumed a 10%
efficiency for the laser, whereas glass lasers then had an efficiency
of about 0.1%. There were signs that it would be hard to deposit the
laser beam's energy on the pellet's corona without also heating the
interior of the pellet and thereby interfering with the process of
cool compression. Various instabilities could also appear that
would interfere with a perfectly symmetric implosion of the pellet.
These were problems that old laser fusion hands like Kidder and his
scientists already knew about and that new players like Nuckolls
and Wood would speedily encounter.[215]

Meanwhile, in 1969, KMS Industries, a private firm, took up laser
fusion energy research. The impetus came from KMS consultant

Keith A. Brueckner, a physics professor at the University of California, San Diego. Brueckner had a background in weapons physics through his consultancies for Department of Defense agencies and defense contractors. He also knew lasers. He had succeeded Townes as vice president and director of research at the Institute for Defense Analyses from 1961 to 1963 and had run a 1963 summer study for IDA on the use of lasers in antimissile defense. He was also familiar with the AEC's magnetic fusion energy program. He had been appointed to the program's Standing Committee in 1968 and in that capacity had attended a fusion energy research conference at Novosibirsk, USSR, in August 1968. There he had heard reports of the Soviet laser fusion experiment and had talked informally to the French investigators.[216]

Since the AEC was declining to initiate a laser fusion program, Brueckner decided to begin one of his own through KMS's Technology Center, where he consulted. By fall 1969, Brueckner and his KMS coworkers had rediscovered the need for compressing the pellet and had independently arrived at the idea of central heating of the pellet. They had calculated that breakeven would only require a few thousand kilojoules of laser energy, a figure not far above the output of the lasers then available. Brueckner concluded that laser fusion energy could be practical. KMS president Keeve M. Siegel took up the idea enthusiastically. He expected the energy industry to grow explosively, and he foresaw that fusion would come to dominate it. He recognized that KMS itself had insufficient funds to carry through a laser fusion program, but he believed that early experimental successes would attract further funding. KMS Industries would then exercise a commanding role through the patents that Brueckner was filing on its behalf. Siegel, himself an engineer and a former University of Michigan professor, estimated that KMS could bring efficient fusion power on the market within a few years after starting an experimental program. A man of some egotism, he was motivated not only by the monetary rewards he anticipated but also by a desire to play a central role in the solution of the nation's energy problems.[217]

KMS awakened vehement opposition within the AEC, where some accused Brueckner of profiting from ideas he had picked up in the course of his AEC and DoD consultancies. By 1970, however, KMS Industries had obtained from the AEC permission to proceed, and by 1972 the company had experimental work under way at an

Ann Arbor, Michigan, facility, using a French neodymium-doped glass laser.

Other U.S. groups were also moving into the laser fusion field. At the University of Rochester, plasma physicist Moshe J. Lubin had been using lasers since the mid-1960s to produce the energetic plasmas that he was studying. By 1968, he had begun to collaborate with scientists in the magnetic fusion energy program to make laser-created plasmas for confinement in magnetic fusion devices. By 1969, he was irradiating deuterium microspheres in laser fusion experiments, and in 1970 he founded the University of Rochester Laboratory for Laser Energetics, an interdepartmental center for advanced study of laser fusion and laser-matter interactions.[218] At the AEC's other major weapons laboratory, the Los Alamos Scientific Laboratory, physicist Keith Boyer started a laser fusion program in 1969. Boyer choose carbon dioxide lasers, which were then just being successfully mode-locked to give the short pulses that fusion demanded. Carbon dioxide lasers had efficiencies closer to those a practical reactor would require. They were also different from the neodymium-doped glass lasers Livermore was using, so that the program would qualify before AEC management as an alternate approach, rather than a mere duplication of the Livermore program.[219] The AEC/Sandia Weapons Laboratory in nearby Albuquerque also began a small experimental program to duplicate the Soviet results.[220]

A small community of U.S. groups active in laser fusion had thus come into being by the early 1970s. The main force that had led scientists into laser fusion, the social demand for pollution-free energy, now began to lead the federal government to increase funding for it. In 1972, the Livermore program was expanded fourfold, from an annual budget of $1 million to one of $4 million. The program was partially redirected from weapons to energy, and a new manager, John L. Emmett from the Naval Research Laboratory, was brought in to replace Ray Kidder.[221] That same year, the Laboratory for Laser Energetics inaugurated an ambitious Laser Fusion Feasibility Project and began to bring industrial firms and government agencies into its organizational structure.[222] By fiscal year 1972, the federal government was spending $20 million a year on laser fusion. By fiscal year 1975, spending was to top $65 million.[223] Laser fusion could not help looking attractive to the scientist as citizen, and as promoter, in the early 1970s.

***

By the time the laser was 10 years old, it was a candidate for a far wider group of applications than the maser had been at its tenth birthday. There were a variety of "plug-in" uses for which incoherent light had previously been employed. These included interferometry, spectroscopy, retinal photocoagulation, and photochemistry. The laser was also being considered for a number of applications, such as communications and radar, for which coherent electromagnetic radiation at longer wavelengths had previously been used. It was being applied in technologies such as machining where its ability to provide high densities of energy was the crucial factor. And there were still other applications that defied these three categories, such as the use of laser beams for alignment in construction projects.

There were, correspondingly, a large cast of protagonists in the drama of the translation of lasers into the workplace. We have seen in these pages many of the types of people that it took. There were academic-entrepreneurs like Moshe Lubin, scientist-managers like Keeve M. Siegel and Arthur Kantrowitz, small entrepreneurial groups within large firms, like Westinghouse's Scientific Technologies Division, and university research leaders like Arthur Schawlow at Stanford. There were government scientists like Kidder, Nuckolls, and Wood, groups studying innovative production techniques, like those at AT&T's Western Electric Engineering Research Center, military R&D agencies that oversee contract research, like the Office of Naval Research, and military laboratories like the Army Missile Command Laboratories or the Air Force Weapons Laboratory.

The diversity of applications and of decision-makers makes it more difficult to generalize about the way the laser passed into the marketplace in the United States than it was for the maser. Some factors in this passage are nevertheless clear. One was the technical state of the art, which was improving almost monthly during the 1960s. A second was the changing pattern of federal expenditures over the decade. This, in turn, consisted both of a tightening of the money available for R&D and a reorientation of federal priorities. And a third—less important but still effective—was national opinion on what were socially desirable technologies. Here consensus was changing with particular rapidity, and a people that began the decade by seeing missiles as a social good ended it concerned with the environment, the plight of urban neighborhoods, and energy.

6

Explaining the Laser

Laser scientists reflecting on the history of their subject often point out that the conceptual ingredients for the invention of a laser were known by the 1930s.[1] The concept of stimulated emission had been introduced by Einstein in the 1910s, and it was understood by the 1930s that it could lead to negative absorption, that is, to amplification. The Fabry-Perot etalon was already being used for the measurement of wavelengths, and some of the methods subsequently needed to achieve population inversions in lasers were also at hand. Why, these scientists have asked, were lasers not invented in the 1930s? Why, in particular, was the optical maser not invented before, rather than after, the microwave maser?

The two known adumbrations of the laser idea that date to the 1930s may be just the kind of exceptions that buttress the rule. The Swiss scientist Fritz G. Houtermans was stimulated by Schawlow and Townes's and Prokhorov's 1958 publications to lay before the scientific public in 1960 an idea he had first developed in 1932— that transient population inversions arise naturally in certain kinds of gas discharges, and that light introduced into one end of such a discharge will emerge from the other end with increased intensity. The Soviet scientist Valentin A. Fabrikant wrote a thesis on the optical properties of gas discharges in 1939 in which he pointed out that it should be possible to create a population inversion in a discharge by adding molecules suited to deplete, by collisions of the second kind, the population of what we would now call the lower laser level. Fabrikant argued that this would allow a demonstration of negative absorption. But Houtermans, although he discussed his idea with colleagues, did not try to reduce it to practice, or even to publish it, and Fabrikant did not mount an experimental program on this line of work until the early 1950s.[2]

Laser physicists, in explaining this situation, have contributed valuable insights into the historical and conceptual structure of quantum electronics. Irina M. Dunskaya has maintained that the concept of stimulated emission was not fully understood in the earlier period. She illustrated her contention by citing a 1928 paper in which two distinguished physicists, C. V. Raman and K. S. Krishnan, interpreted radiation from a system of molecules with a Boltzmann (that is normal, rather than inverted) population distribution as stimulated emission.[3] Mario Bertolotti has observed that, although stimulated emission and Fabry-Perots indeed came early, the 1940s and 1950s saw the growth of new fields of physics that were to provide masers and lasers with a rich store of other useful techniques and ideas. One of these fields was nuclear and electron magnetic resonance. As we saw in chapter 2, Joseph Weber drew on Purcell and Pound's nuclear magnetic resonance research in elaborating his maser proposal. Nicolaas Bloembergen brought the full panoply of concepts from nuclear and electron magnetic resonance into his work on the solid-state maser. Another of the new fields of the late 1940s and early 1950s was optical pumping. It was originally introduced, by the French physicist Alfred Kastler and others, as a means for changing the populations of various levels in order to study the properties of excited systems. By the mid-1950s, it was also to become a technology for pumping masers and, shortly thereafter, lasers.[4]

Michael Feld, in an unpublished master's thesis submitted in 1963, made the question of why prewar scientists did not invent the laser the subject of an interesting historical comparison. Feld emphasized the difference in mindset between the physicist of the 1930s, who was preoccupied with verifying and using the newly discovered quantum mechanics, and the physicist of the post–World War II period, for whom quantum mechanics was a settled theory. By the postwar period, also, the idea of the absorption of radiation by a group of molecular systems had been reconceptualized as an idea of a resonance between the frequency of the radiation and the frequency of energy transitions in the molecular systems. "Resonance" is not far from "resonator," and the microwave physicists of the postwar era, who had learned engineering through radar research carried out during the war, were familiar with the latter as well as the former. Feld also emphasized that resonance phenomena were now being studied in solid media, such as lithium fluoride, in which it was far easier to produce a complete population inversion than in the gas discharges studied by the scientific workers of the 1930s.[5]

Charles Townes, in his article "Ideas and Stumbling Blocks in Quantum Electronics," wrote that a number of physicists had toyed with the idea of obtaining amplification not in a technological context, but in a physical one, as a demonstration of the phenomenon of stimulated emission.[6] But the experiments appeared to be difficult, and stimulated emission could be verified by other effects. The crucial stimulus to develop the idea of the maser was the conjoining of the knowledge of the physicist with that of the engineer. "It is sometimes said that there is no single component idea involved in the construction of masers or lasers which had not been known for at least 20 years before the advent of these devices . . . . I believe whatever unnecessary delay occurred was in part because quantum electronics lies between . . . physics and electrical engineering. . . . [T]he necessary quantum mechanical ideas were generally not known or appreciated by electrical engineers, while physicists . . . were often not adequately acquainted with pertinent ideas of electrical engineering." "It is understandable," Townes continued, "that the real growth of this field came shortly after the burst of activity in radio and microwave spectroscopy immediately after World War II since this brought many physicists into the borderland area between quantum mechanics and electrical engineering." Elsewhere, Townes has pointed out that it was wartime radar research that occasioned this "burst of activity" by setting some of the scientific problems, by providing war-surplus equipment for the experiments, and by training physicists in techniques for generating radio and microwave frequencies.[7]

The foregoing analyses have been mainly in terms of events "internal" to physics, that is, they have focused on changes in the concepts, or experimental modalities, available to scientists. Townes and Feld transcend this limitation, but even they bring in external factors, such as World War II, at one remove, through the effect they had on scientists' attitudes and intellectual furniture. This is not surprising. Historians of science and technology have moved only in the last few decades from explaining science in terms of internal intellectual factors to explaining it in terms of a complex commingling of social and technical causes. Contemporary scientists, it is fair to say, are still internalists. That is evident in the very fact that there is currency for a question like: Why was the laser not invented in the 1930s? Such a question presumes that the crucial precondition for a scientific novelty is the stock of concepts available to the community of practitioners. What the contemporary historian will want to add to the insights of scientists is, above all, a social dimension.

One procedure that allows us to place the social dimension on a parity with the technical one is to ask about the groups that inaugurated laser research: What were the properties of this research that made it attractive to them? Since hundreds of U.S. research teams took up lasers in the 1960s, we have, at least potentially, hundreds of case studies to draw upon in answering this question.

The preceding chapters in effect provide a handful out of these hundreds of cases. They show that bench scientists in universities, industry, and government laboratories took up lasers for cognitive reasons, to be sure, but also for economic, social, and even psychological ones. The different types of motivations were so entangled that they are probably impossible to unscramble.

We have seen that bench scientists were intrigued by the scientific questions that the operation of the first lasers laid on the table. Could media other than Javan's helium-neon mixture or Maiman's pink ruby or the doped calcium fluoride crystals of Stevenson and Sorokin be made to lase? Were excitation methods other than optical pumping or collisions of the second kind feasible? How should the coherence of laser light be analyzed? What new effects could be created in inorganic or biological materials by the laser beam's enormous electric fields? What tasks in experimental physics, chemistry, and biology could the laser help to perform?

The economic and social motivations that have to be added to these intellectual inducements are multifold. In the first instance, there was the money that the military was putting up for laser research and development. Then there was the social circumstance that the laser marked out a good route to professional recognition. This partly devolved from the fact that lasers offered a virgin field, in which new, important, and eminently publishable results lay around for the picking. Finally there were emotional factors. The sheer excitement of participating in an area that was so much in the eye of the scientific community and the public was a reward in itself, as was the exhilaration of the rapid rate at which new discoveries were emerging.

But it is not only bench scientists for whom we need to account. The laser interested many other actors in the U.S. research establishment, including high-tech entrepreneurs, industrial managers, military R&D monitors, and officers of scientific societies. The question of why the laser drew their attention entails a further question: What was the reward system within which these individuals and groups operated, and which of the rewards available to them did the laser promise? Again, it will be the aggregate of laser case studies that will furnish material for an answer. In effect, one identifies segments of the research establishment modulo laser and exposes its award system modulo laser. Looking at other branches of science and technology in the same epoch will give a somewhat different portrait of U.S. R&D; looking at other epochs will give a sense of the research establishment's change with time.

These ideas can perhaps be made more useful in the form of a chart. In table 6.1, column 1 lists the sectors that participated in laser research, column 2 gives some elements of the reward system that have been revealed by the case studies in this book, and column 3 tells us what might have seemed to be "in it for them" as members of these sectors looked at the laser. The laser became a "hot topic" precisely because its properties were well matched to the reward structure in a large number of sectors of the U.S. high-technology establishment as it was constituted in the early 1960s.[8]

Missing from the table are the connections that bound the sectors together, connections that we have seen repeatedly throughout this book. Academic scientists were linked to industrial scientists through the consultancies that university professors held in large and small firms, through the industrial sponsorship of university fellowships, and through the placement of university graduates and postdoctoral fellows in industry. They were linked by joint projects, of which a major example here is the Townes-Schawlow paper of optical masers, and through sabbaticals that academics took in industry and industrial scientists took in universities. Academic scientists were linked with the Department of Defense R&D groups, and with other government agencies through tours of duty in research organizations such as the Institute for Defense Analyses, through work at DoD-funded laboratories such as the Columbia Radiation Laboratory or the MIT Research Laboratory for Electronics, and through government study groups and consultancies. They were also linked by the fact that so much of their research was supported by the Department of Defense and NASA.

Large industrial R&D laboratories were linked with small laser firms through the small firms' reliance on the research results of

Table 6.1
The U.S. Research Establishment and the Laser, 1961

Sector	Reward structure	What the laser offered
Professional societies	Growth of membership. Legitimacy within scientific community. Financial resources. Growth of field served. Vigor of publications and meetings.	Linkage of optics and engineering to basic physics. Access to new pools of potential members. Government support for conferences, etc. Large pool of research papers for meetings and journals. Support from new sectors of industry.
Academic scientists	Professional and public recognition. Joy of interesting and important[b] science. Resources for doing science and for training students.	Intellectual challenges. Publishable results that commanded attention. Government contracts and consultantships. Jobs for students. New instruments for doing science.
Industrial scientists	Joy of interesting and important science. Recognition within discipline. Recognition and advance within company. Resources for doing science.	Intellectually challenging projects. Publishable results that commanded attention. Contributions to company technology. Company and government resources for science.
Industrial R&D managers	Advance of company technology. Enhanced reputation for laboratory. Strong patent positions.	New technologies. Government contracts. Enhancement of laboratory prestige.
Laser firms	Profitable markets. Contract money. Availability of capital. Interesting technology.	Money from government contracts and venture capital. Low barriers to entry. Demand for experimental lasers from academic and industrial scientists. Frontier technology. Expectations of large markets.
Federal R&D agencies and in-house laboratories[a]	Advance of military technology. Advance of scientific disciplines that underlie military technology. Recognition and advance within military. Creation of a clientele of outside researchers.	Creation of large in-house programs. Antiballistic-missile technology. Battlefield aids (radar, smart bombs, communications). Creation of new career tracks.

a. This row relies on the work of R. W. Seidel, "How the Military Responded to the Laser," *Physics Today* (October 1988), 12–19.
b. The qualification of science as "important" is a complex social process.

the industrial giants: Spectra-Physics used Bell Laboratories' helium-neon lasers, for example, while Coherent used their carbon dioxide laser. Small companies also received some of their capitalization from large companies. In return, they provided research and production equipment to the larger firms and at times engaged with them in joint applications research. Professional societies had relations with academic researchers, with industry (through individual and corporate industry memberships), and with DoD agencies, which often funded and cosponsored conferences. Industrial scientists and industrial R&D managers were linked to the government through Department of Defense and NASA contracts and through membership in advisory panels and study groups. This matrix of interconnections was reinforced by the mobility of the laser scientists. We have seen many of them move in these pages from government laboratories and industry to universities, and from one company to another.

These linkages had a profound effect on the consciousness of the scientists. Hugh G. J. Aitken, in *The Continuous Wave*, has described how Heinrich Hertz invented the transmitter and receiver that became the elements of radio for reasons purely internal to the science of physics, to confirm James Clerk Maxwell's electromagnetic theory. Aitken shows how Guglielmo Marconi then served as a translator of Hertz's results into a commercial product, through his vision that Hertz's laboratory technology could be bent to the use of ship-to-shore communications.[9] A postwar U.S. Hertz is almost inconceivable, for academic physicists had acquired triple vision and were themselves looking out for the military and commercial as well as the scientific applications of their inventions. An analogy to markets may be helpful. There were disciplinary markets, military markets, and commercial ones for these devices. U.S. scientists were often well acquainted with the needs of two or more of these markets. I would argue that the fact that many scientists, in whatever organization they were located, had all these needs in mind probably raised their level of effort. That is, scientists were more likely to pursue work they might be able to "sell" in more than one market.

The linkages between U.S. research sectors in the post–World War II period have been the subject of much recent historical analysis.[10] It is fair to say that there has been some division of labor among historians dictated by their subject matter. Studies of U.S. accelerator physics, for example, have focused attention on the relation of university scientists to government agencies, while studies of semiconductor technology have explored the nexus between government and industry and between large industrial laboratories and small high-technology firms. I would suggest that the virtue of the laser case is precisely that it forces us to attend to all the members of the triumvirate of industry, university, and the military. This is a consequence of the position of the laser on the interface between science and technology, which compels a joining of the perspectives of the historian of physics and the historian of technology. Science-based technology, rather than science itself, will, I believe, prove the most favorable ground for exploring the structure of research in the United States.

Epilogue: The Laser Now and in the Future

A. H. Guenther, H. R. Kressel, and W. F. Krupke

Introduction

During the 1980s, the practical importance of lasers grew markedly as their properties became better understood, their costs decreased, and their reliability improved. This growth was supported by the parallel development of electronic and optical adjuncts that allowed lasers to be employed effectively in a variety of applications. The development of additional types of lasers extended their utility: Lasers with emission wavelengths ranging from x-rays to the far infrared are now available, with greatly differing power levels, efficiencies, sizes, operating conditions, and temporal output characteristics.

In the early years, enthusiasm for lasers got in the way of good business sense. The laser technologist (acting as salesman) tended to promote the laser for its own sake, rather than as part of a system or instrument. Laser manufacturers paid inadequate attention to market size: Although lasers provided unique solutions to some industrial problems, the demand was all too often insufficient to underwrite the cost of product development or yield adequate profit. Industrial users, comfortable with the status quo, showed a natural reluctance to accept the new approaches the laser offered.

As time went on, however, the advantages of lasers became apparent as major applications arose in industry, government, and medicine. In 1987, the annual market value of lasers, worldwide, was estimated to be about $600 million. The total annual value of systems containing lasers was estimated to exceed $5 billion. The number of lasers varies with type: each year, large quantities of low-power, semiconductor laser diodes and helium-neon lasers are manufactured, but only a few thousand Nd:YAG, carbon dioxide, and argon ion lasers of moderate power are produced.

In this epilogue, we will review some of the more important present-day applications, grouped under seven headings: fiber-optic communications, industrial applications, consumer and computer products, scientific and industrial instrumentation, military applications, energy applications, and medical applications.

Fiber-Optic Communications

The conjunction of semiconductor laser diodes and optical fibers has transformed the telecommunications industry. Digitized information, in the form of light produced by the laser diodes, can be propagated over optical fibers with unprecedented economy. The first commercial digital communications systems were deployed in 1976, operating at 45 megabits per second over a few kilometers. Since then, fiber-optic transmission systems have largely replaced copper-wire-based systems for long-distance communications, and the growth rate of systems using microwaves and satellites has sharply declined. The dominance of fiber-optic communications is based on economics: Such systems provide the lowest cost and highest reliability for high-bandwidth transmission, both on land and under the sea.

Commercial fiber-optic communication was made possible by two major innovations around 1970: laser diodes capable of continuous operation at room temperature and glass fibers with the relatively low attenuation of 20 decibels per kilometer at 8500 Å, the same wavelength as the emission from the aluminum-doped gallium arsenide laser diode.

Laser Diodes: The Early Years

The first semiconductor laser diode, demonstrated in 1962, created enormous excitement even though it initially operated only at or below the temperature of liquid nitrogen. Here was a laser the size of a grain of sand that potentially could replace much larger, less efficient lasers! The ensuing applied research focused on devices using gallium arsenide (GaAs) and its alloys, particularly GaAsP. Experimental lasers emitting from 7000 to 9000 Å could be made by changing their phosphorus content.

Between 1962 and 1966, work on practical laser diodes encountered formidable technological obstacles: (1) Threshold current densities increased very steeply with temperature, exceeding 100,000 amperes/centimeter^2 at room temperature. These current densities made thermal dissipation an enormous problem. (2) Operating lifetimes were erratic. Lasers functioned for periods of time ranging from seconds to hours, depending on the sample, operating conditions, and uncorrelated process variables. (3) Power conversion efficiencies were low and decreased with increasing temperature. As a result, the lasers emitted at low levels of power, were unreliable, and therefore were of little practical interest.

These daunting problems dampened enthusiasm for the laser diode as a commercially useful device. After 1966, significant efforts were being made at only a few industrial laboratories. In the United States, these included RCA, IBM, Bell Telephone Laboratories, and Lincoln Laboratory.

Development of the laser diode continued under the auspices of the Department of Defense, which perceived such military uses for the technology as line-of-sight communication over land and sea and infrared illuminators for night vision. (The latter required arrays of GaAs laser diodes enclosed in refrigerated units designed to cool the devices to liquid nitrogen temperature.) Little progress was made toward commercially viable applications until this state of affairs was profoundly changed by the invention of the heterojunction laser diode.

The Heterojunction Laser Diode

The emergence of the modern semiconductor laser diode dates from the invention of the heterojunction laser diode, a multilayer structure formed by the sequential deposition of thin, semiconductor films having carefully matched, differing band-gap energies. The first devices, which are still the most widely used, consisted of layers of GaAs and AlGaAs.

The development of the single heterojunction laser was completed by RCA in 1968, and it was offered on the commercial market in 1969. The device was the first to have a sufficiently low threshold current density at room temperature (10,000–15,000 amperes/centimeter^2) to allow reliable emission of pulsed peak power in the tens of watts from a single diode.

The double heterojunction laser, developed by laboratories in the Soviet Union and in the United States, provided a much lower threshold current, permitting continuous-wave (CW) operation at room temperature. The 4,000 amperes/centimeter^2 current densities achieved in 1970 were reduced as low as 200 amperes/centimeter^2 by 1987 as a result of continuing refinements. It consequently became possible to modulate the output of the laser diode at any rate up to a frequency limit beyond 1 gigahertz, making it a practical light source for optical communications.

While the heterojunction laser solved the problem of operating power level, the reliability of the early devices remained inadequate. Here, progress came through identification of the failure mechanisms and improvement of both the structure and the production processes. By 1980, the predicted lifetimes of AlGaAs lasers at room temperature exceeded 50,000–100,000 hours. Such values were sufficient for many practical systems, including fiber-optic communications, and device reliability ceased to be a major limitation to the deployment of laser diodes.

The GaAs/AlGaAs laser diode has established itself as the most widely used lasing device, with a total of nearly 12 million units produced in 1987. As the production volume of laser diodes increased, economies of scale made possible reductions in unit price. Certain AlGaAs devices used in consumer products have declined in cost from hundreds of dollars to less than five dollars. This, in turn, has further broadened their use.

Second in importance to the AlGaAs lasers are devices containing InGaAs and InP structures, which emit in the 1300–1500 nanometer region of the spectrum (a nanometer is 10^{-9} meter or 10 Å). In 1976, fused silica fibers were produced that had transmission losses of only 0.5 decibel/kilometer in this region. Furthermore, pulse spreading in these fibers was minimal at about 1300 nanometers. Since lasers emitting at this wavelength could not be made with the technology then available, lasers based on the new alloys InGaAsP and InP were created to fill the need.

By 1980, commercial fiber-optic systems using InGaAsP lasers and single-mode fibers began to be deployed. They operated at 565 megabits per second over distances of tens of kilometers. By 1987, the commercial system capability had been extended to 1.3 gigabits per second, and higher rates are expected in the future. Fiber-optic links now span the United States, Europe, and Japan and are being incorporated in all new land and undersea telecommunications systems worldwide.

Industrial Applications

The major industrial uses for lasers are in materials processing (cutting, welding, heat treating, and marking) and in instruments used for measurement, control, or characterization. The power levels of laser output range from a few milliwatts for marking to as high as 10 kilowatts for cutting thick metals. A unique asset of the laser is that this highly controlled beam of energy can be directed with micrometer precision to heat a very well controlled area intensely.

Two main types of lasers are used for materials processing, Nd:YAG and carbon dioxide, with the choice determined primarily by the properties of the material being worked. High-power carbon dioxide lasers have made possible machine tools that can cut materials with unprecedented speed and precision, creating intricate patterns in materials ranging from steel to cloth without fraying or cutting-tool degradation. Cutting accounts for over 60% of the use of such lasers in industry. Nd:YAG lasers have two main uses: trimming resistors to obtain the precisely controlled resistance values required for integrated circuits (accounting for about 40% of their use), and as components of welding equipment (about 30% of their use).

Although laser-based systems are displacing conventional materials-processing technology relatively slowly, certain industrial processes are uniquely dependent on laser-based technology. For example, in the microelectronics industry, lasers remove excessive pattern material, in micron-size features, from the photo masks used to fabricate integrated circuits. Without such tools to repair defects in masks, economical manufacture of the sophisticated memories and logic devices essential to modern electronics and computers would be impossible.

Consumer and Computer Products

The laser's ability to project intense light to spots as small as one micrometer in diameter is widely used in encoding and reading data. Low-power helium-neon lasers are commonly used in retail stores to scan bar codes. Diode lasers are employed in inventory and process control and to detect the information stored in prerecorded digital form on audio or video discs. With over 10 million lasers sold in 1987 for compact disc playback, this is now the largest market for laser diodes.

In computer data storage systems, the laser diode is used to encode and read digital data on discs in the form of spots about 1 micrometer in diameter produced by local heating. The reflection of the beam from the spot creates readable information. Technology exists for both permanent and erasable discs. Optical discs provide much greater storage capacity than conventional magnetic discs and are becoming increasingly important for storing computer data.

Printers attached to computers increasingly rely on laser diodes for encoding information on photosensitive surfaces that then transfer it to paper. These laser printers are much faster than the mechanical printers they replace and will eventually yield book-quality printed material. They are becoming the standard printers for office use, since their product is much superior to that of a typewriter, dot-matrix printer, or (in the case of math symbols) handwriting. About 1.4 million laser diodes were used in laser printers in 1987.

In some simpler forms of data encoding, lasers are used to "burn" information into various surfaces. The type of laser employed depends on the material to be marked.

Scientific and Industrial Instrumentation

Scientists and engineers, the people who knew the laser best, were the first to perceive and exploit its unique capabilities. They grasped it as a means to probe phenomena that were hitherto inaccessible and to obtain fundamentally new scientific and engineering information. The results of this laser-aided research stimulated the development of new lasers and laser applications.

Progress in laser technology, indeed, has been autocatalytic, to borrow a concept from the chemists. For example, the laser demanded higher-quality optics, but it also became the preferred light source for testing and measuring the quality of optical components. The laser, for example, revealed an unsuspected roughness on the surfaces of conventional optical components that gave rise to scatter loss. Fabrication techniques were subsequently developed that could yield optical surfaces with low scatter loss. Another example of how laser development fed on itself is laser-induced fluorescence, which bred new methods for studying the kinetics of chemical reactions, some as basic as combustion, while uncovering possible laser materials as the kinetics of certain reactive systems were unraveled.

As measurement tools, lasers have made their greatest impact in all types of spectroscopy. Dye lasers are the most versatile spectroscopic adjuncts because their output radiation can be tuned in frequency. Semiconductor lasers are becoming steadily more useful, as are the color-center laser and the emerging hollow-cathode devices, which operate in the ultraviolet. Dye lasers can maintain their stability and spectral purity while operating continuously. They can also produce repeated pulses of light intense enough to induce nonlinear processes such as multiphoton absorption or frequency conversion. Exciting vistas have been opened by lasers producing single, short pulses (coherent optical transients) for studying coherent photon echoes from atoms.

Lasers contribute unique capabilities to many conventional instruments. The use of such laser-enhanced instruments ranges from materials analysis to building construction; for example, lasers are used to vaporize small samples in mass spectrum analyzers and to induce fluorescence by selective excitation. Visible laser beams are used with construction equipment to control ground leveling, pipe laying, and tunnel boring. As measurement instruments, lasers are employed in geodesy (surveying) and to control the position of machine tools (figure E.1).

Military Applications

Both high- and low-power lasers are part of the national armamentarium. Optical radar and laser weapons require high-power lasers, while low-power lasers figure in rangefinding, target designation, and proximity fusing.

High-Power Military Applications

The United States has spent in excess of a $1 billion over the past 15 years in an effort to develop efficient, powerful lasers usable as directed-energy weapons. Such weapons could be used to destroy or defend satellites. Another envisioned application, the most demanding of all, is the interception of incoming ballistic missiles. The basic appeal of lasers as weapons derives from the fact that they can deliver a lethal dose of energy to a target at the speed of light and can therefore take on missions with very stressing time lines.

These applications favor lasers with short wavelengths—visible, ultraviolet, or x-ray—because their effectiveness is ultimately constrained by a fundamental property of light: diffraction. Coherent light energy at a wavelength λ can be projected from an aperture of

Figure E.1
Laser leveler in use on a construction project. (Courtesy Spectra-Physics, Construction and Agricultural Division)

diameter D to a distant target with a minimum spot size d that varies as λ/D. Thus, for a given amount of laser energy or power and a given target vulnerability and susceptibility to radiation, the shorter the wavelength, the smaller the diameter needed in the optical projection system. Furthermore, the fact that d is proportional to λ/D tells us that for an aperture of given diameter, if we decrease the wavelength, targets at a greater range can be effectively intercepted. To provide lasers of short wavelength for directed-energy weapons, exploratory research and development was initiated in the early 1970s at university, industrial, and government research laboratories. Three distinct types of short-wavelength lasers came out of these fruitful efforts—chemical, rare-gas halogen, and free electron—all with the potential to be scaled to high energy and power.

Initial research efforts centered on finding gas-phase chemical reactions that would produce excited molecules with gain on visible wavelength transitions between *electronic* states. At first, the most promising candidates for such visible chemical lasers were the Group IV-A fluorides, typified by SiF, GeF, and SnF, and later by the diatomic molecules IF, SO, and NO. Unfortunately, this line of attack has so far proven disappointing because the radiative lifetime of an upper laser level decreases as the cube of the wavelength. Therefore, as the wavelength is shortened, the excitation (chemical reaction) rate must be proportionally increased to maintain a sufficient population inversion density. A variety of chemical laser schemes were evaluated, but none seemed capable of generating the high density of reactants and the large gas-mixing rates that were needed.

An alternative, successful strategy for generating laser light chemically was proposed and demonstrated in the late 1970s. Electronically excited oxygen molecules are prepared chemically (in sufficient density) and flow-mixed with iodine vapor. A complex series of chemical and kinetic reactions then produce laser emission at 1315 nanometers. The chemical oxygen iodine laser has subsequently been scaled above the 25-kilowatt level and is under development for potential use as an antisatellite weapon.

Another type of "chemical" laser, the rare-gas halogen excimer laser, was demonstrated in 1975. In its simplest form, a rare gas such as krypton is premixed with a gas such as fluorine. An electrical discharge is passed through the gas mix, producing halogen ions that react chemically with the electronically excited rare-gas ions to produce electronically excited rare-gas/halogen molecules. These molecules relax by emitting a photon in the near-ultraviolet spectral region and dissociating into free atoms. Initial results, particularly with the xenon fluoride excimer (XeF) operating at 353 nanometers and the krypton fluoride excimer (KrF) operating at 249 nanometers, triggered intense excimer development efforts. These studies indicated that these lasers could be scaled in energy and power to levels of interest for directed-energy applications, with power efficiencies in excess of several percent. Following President Ronald Reagan's March 1983 speech calling for the development of a system for defense against strategic missiles (the Strategic Defense Initiative), excimer lasers were proposed as directed-energy sources for defense against ballistic missiles. But by 1983, excimer lasers were no longer the only powerful, short-wavelength lasers available; the extremely promising free-electron laser (FEL) had appeared on the scene (figure E.2).

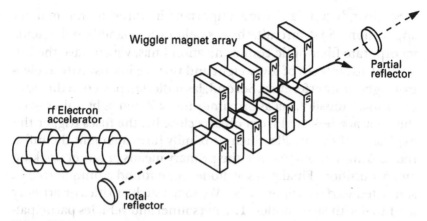

Figure E.2
Diagram of a free-electron laser. In his description accompanying the original publication, Charles A. Brau notes that "the beam of electrons . . . forms laser radiation." (Adapted from Brau, "Free Electron Lasers," *Science* 239 (4 March 1988), 1116)

The free-electron laser exploits the coherent transfer of kinetic energy from a narrow beam of relativistic electrons to a copropagating beam of electromagnetic radiation in the presence of a periodic, static magnetic field (a "wiggler"). Grounded in 1950s work on backward-wave electron tubes, the modern laser version was conceived in the early 1970s at Stanford University. Free-electron amplifier gain was first demonstrated at a wavelength of 10,600 nanometers in 1976, followed by a demonstration of laser oscillation at 3400 nanometers in 1977. In contrast to conventional laser media, for which the wavelength is fixed by atomic or molecular parameters, the peak wavelengths of the FEL gain can be tuned by varying the voltage of the electron accelerator or by varying the spatial period of the magnetic field. In addition to having this novel feature, the FEL has a relatively high efficiency (tens of percent), good beam quality, and an output power expected to be scalable well into the megawatt region.

Low-Power Military Applications
The laser is the basic element of the rangefinder that is in general use in the U.S. Armed Forces. The ruby laser originally used for this purpose gave way to the Nd:YAG laser in the 1960s. Low-power Nd:YAG lasers are also used in target designation for missile guidance.

Semiconductor lasers are important in three newer military applications. First, aircraft, ships, satellites, and land-based systems incorporate fiber-optic communication links, which have the important advantages of low weight and relative insensitivity to electromagnetic interference. Second, laser diodes, placed in the sides of air-to-air missiles, act as proximity fuses: When light pulses from the laser are reflected from objects close by, the fuses trigger the explosion of the missile. Such proximity fuses augment the infrared-seeking sensors that are the primary means of target tracking and acquisition. Finally, laser diodes have found a unique role in simulated warfare (figure E.3). Weapons such as rifles or artillery are fitted with laser diodes. The personnel and vehicles participating in these war games carry light sensors designed to sense the coded radiation from the lasers. A "hit" consists of the registration by a sensor of the laser light pulse. This system has proven far superior to the subjective evaluation methods previously used in such training exercises.

Energy Applications

Laser Fusion

High-power lasers can serve as the energy source for inducing laboratory-scale thermonuclear reactions. In laser-driven fusion, a spherical target containing a thermonuclear fuel, such as a mixture of deuterium and tritium, is uniformly irradiated by a brief laser pulse (less than 10 nanoseconds). The target implodes, forming a plasma dense and hot enough to cause a thermonuclear reaction. If a hundredfold multiplication of energy can be gained, for an input of laser energy of a few megajoules or less, significant military and civilian applications are possible. Such high-gain fusion reactions could serve as laboratory experiments in the physics of nuclear weapons and would aid in the design of such weapons. In addition, laboratory-based fusion would permit scientists to study the effects that nuclear weapons would have on such military targets as satellites, reentry vehicles, and electronic components.

The civilian application of laboratory-scale fusion is the generation of electricity in central power stations. This will be feasible if a hundredfold energy gain can be achieved at repetition rates of about 10 pulses per second (10 hertz) at a laser efficiency greater than 10% and a cost less than $50 per watt.

Figure E.3
This early infantry weapons simulation system was developed by International
Laser Systems in the mid-1970s. (Courtesy Schwartz Electro-Optics, Inc.)

In the late 1960s, a community of scientists from industry, government, and universities interested in laser-driven fusion formed in this country and abroad. Simple theories of the interaction between laser light and plasmas indicated that lasers of shorter wavelength would better avoid the problem of plasma instabilities, but the question was how short a wavelength is short enough to drive a high-gain (i.e., roughly hundredfold) implosion. At that time, the lasers judged to be scalable to the megajoule/10-nanosecond regime were carbon dioxide, at 10,000 nanometers, HF at 2700 nanometers, atomic iodine at 1300 nanometers, and neodymium-doped glass at 1060 nanometers.

Neodymium-doped glass had the advantage in wavelength. Furthermore, were 1060 nanometers to prove an insufficiently short wavelength, it was anticipated that the radiation from a neodymium-doped glass laser could be effectively shifted to still shorter wavelengths—green at 530 nanometers, blue at 353 nanometers, and ultraviolet at 266 nanometers—using nonlinear crystals as harmonic converters. In addition, this laser was capable of providing the range of pulse durations and shapes that would be necessary in an experimental facility intended to explore a wide variety of fusion target designs. But while neodymium-doped glass was suitable for the nearer-term military applications, its low efficiency (less than a few percent) and its poor potential for scaling up to high average power limited its use in the longer-term application of commercial power generation. The carbon dioxide laser, on the contrary, had the potential for high efficiency and high average power, but it was handicapped by its relatively long wavelength.

Within the U.S. federal laser fusion program, the Los Alamos National Laboratory chose the high-efficiency carbon dioxide laser approach, while the Lawrence Livermore National Laboratory chose the "short-wavelength" approach. Livermore embarked on two development programs: neodymium-doped glass lasers, fitted with harmonic converters if necessary, to obtain shorter wavelength emission, and entirely new short-wavelength lasers with the efficiency, average power, and cost required for fusion-generated power production. The neodymium-doped glass laser was also chosen for fusion research by most of the other fusion research laboratories in the United States (University of Rochester, KMS Fusion, Naval Research Laboratory), the Soviet Union (Lebedev Physical Institute), Europe (Limeil in France and the Atomic

Weapons Research Establishment in England), and Japan (Osaka University). Significant efforts were made to develop carbon dioxide fusion lasers at the Lebedev Institute and Osaka University and a fusion iodine laser at the Max Planck Institute at Garching, Germany.

A series of short-pulse (about 1-nanosecond) carbon dioxide laser systems with increasing energy and power were designed and constructed at Los Alamos between 1974 and 1984. The largest of these, ANTARES, produced an energy of over 30 kilojoules in 1-nanosecond pulses. Fusion experiments performed with this laser persuaded researchers that such a long wavelength produced plasma instabilities too severe to overcome. By 1984, active pursuit of carbon dioxide lasers for laser fusion was generally abandoned.

A series of short-pulse neodymium-doped glass laser systems of increasing energy and power were designed and constructed during the 1970s and 1980s. Earlier lasers operated at the fundamental wavelength of 1060 nanometers, whereas later lasers were operated at one or more of the wavelengths of 530, 353, and 266 nanometers. The development of neodymium-doped glass fusion laser systems culminated in the 100-kilojoule class Nova laser system at Livermore in 1985 (figure E.4). The worldwide, decade-long use of neodymium-doped glass laser systems for fusion-target experiments has established the fact that short-wavelength radiation does have a beneficial effect in reducing deleterious plasma instabilities. Present estimates indicate that a high-gain implosion can be achieved at a laser energy less than 10 megajoules for wavelengths under 350 nanometers.

Research and development of laser systems suitable for driving a fusion reactor power plant are in progress. Los Alamos and several other laboratories in Japan and Europe are developing KrF excimer lasers, while Livermore is developing free-electron lasers and gas-cooled solid-state lasers pumped by efficient semiconductor laser diode arrays.

Laser Isotope Separation

A second use for lasers in energy production is to enrich natural uranium from 0.7% U^{235} to 3% U^{235}, making it suitable as a fuel for light-water nuclear reactors. AVCO Everett Research Laboratories demonstrated uranium enrichment on a laboratory scale in 1971. In their ensuing joint venture with Exxon, they demonstrated an industrial-scale process in 1976 and began plans for a pilot plant.

Figure E.4
The Nova target chamber, 1985. (Courtesy Lawrence Livermore National Laboratory)

Their process utilized atomic uranium vapor as the working material. Government-financed programs at Lawrence Livermore National Laboratory (using atomic vapor) and Los Alamos National Laboratory (using molecular uranium hexafluoride) were also started in the early 1970s. By 1980, however, although military demand for enriched uranium and plutonium had increased, commercial prospects had declined dramatically. Exxon pulled out of its joint venture with AVCO in 1980, after a total expenditure of more than $75 million. In 1985, the U.S. Department of Energy selected the Livermore atomic vapor process as the future technology for uranium enrichment capacity in the United States. An analogous atomic vapor process for the purification of plutonium for military applications was also selected by the Department of Energy.

Medical Applications

The application of lasers to surgery was envisioned immediately following their invention, for several reasons. A laser could deliver a substantial amount of energy to a small volume of tissue, thus heating and destroying it through a number of mechanisms. The laser's ability to localize irradiation precisely would make it ideal for microsurgical operations, and the absence of physical contact with the subject tissue would promote sterility. In addition, the projection of laser radiation through optical fibers could allow noninvasive procedures to be performed deep within the body.

The degree of heating of tissue depends on its absorption characteristics, on the laser's wavelength and intensity, and on the duration of exposure. Blood hemoglobin absorbs strongly at wavelengths shorter than about 600 nanometers and is relatively transparent well into the middle infrared region. In contrast, water, which makes up about 90% of the mass of all body cells, is relatively transparent throughout the visible and near infrared but absorbs strongly beyond about 2000 nanometers. Thus, a wide range of laser-mediated effects is possible.

Radiation from the argon ion laser, with wavelengths in the 488–512 nm region, is heavily absorbed by blood hemoglobin, rupturing blood cells, activating platelets, and damaging the lining of blood vessels. This sequence of events promotes blood coagulation, which produces clotting in the vessel and cessation of bleeding. The depth of penetration in tissue is on the order of a millimeter.

In contrast, the 10,600-nanometer radiation from a carbon dioxide laser is strongly absorbed in cells predominantly composed of water, and this laser is consequently favored for vaporizing and cutting tissue. The intense heat caused by absorbing laser radiation generates steam, which ruptures the cell. Impact depths of about 0.1 millimeter are typical, with injury to surrounding tissue being limited to about 100 microns. These factors also render the carbon dioxide laser appropriate for the removal of nonvascular tissue, such as brain tumors, and connective tissue. Recently, solid-state erbium lasers operating at wavelengths near 2900 nanometers (matching the spectral region of the strong onset of absorption by water) have begun to be evaluated for many of the procedures now performed by the carbon dioxide laser.

The 1060-nanometer radiation of the Nd:YAG laser is not heavily absorbed by blood or water. Its radiation can pass through blood clots and coagulate larger vessels not well treated with argon laser radiation. The high peak-power level generated by pulsed Nd:YAG lasers finds application through the mechanism of photodisruption. Here the intense optical field induces absorption in normally transparent tissue through nonlinear (multiple-photon) processes and leads to the formation of a plasma. Rapid plasma formation, in turn, creates intense, hypersonic shock waves, which disrupt surrounding tissue.

In the past few years, it has been found that pulsed excimer gas lasers emitting in the ultraviolet region can produce exceptionally clear and precise etching in organic matter on a subcellular scale with little or no thermal damage. The ultraviolet photon energy is high enough to break molecular bonds directly (rather than through the usual thermal mechanism), with excess energy carried away by ablated molecules in the form of vibrational and translational energy.

Historically, it was in the area of ophthalmology that lasers were first intensively and successfully used. Photocoagulation experiments to treat a variety of retinal disorders began as early as 1961 with a pulsed ruby laser, followed in the mid-1960s by the development of techniques based on the continuous-wave argon ion laser emitting in the visible. Today ophthalmic lasers are routinely used to treat the leading causes of blindness: age-related macular degeneration, diabetic retinopathy, and glaucoma. In panretinal photocoagulation, 1,000 to 3,000 laser burns about 500 microns in size are created in the retinal periphery with a CW argon laser. This treatment induces regression of the abnormal blood cells, decreasing the chance of hemorrhage and retinal detachment. Glaucoma (elevated intraocular pressure) can be corrected by a laser-based iridectomy in which the flow of aqueous fluid from the posterior chamber of the eye to the anterior chamber is returned to normal by widening the opening in the iris. Argon ion lasers are widely used for this procedure, but recently the photodisruptive mechanism achieved with pulsed Nd:YAG lasers has been employed. Corneal sculpting (radial keratonomy) with pulsed excimer lasers is being developed as a means of correcting the eye's ability to focus by altering the surface shape of the cornea.

Many other fields of surgery have also found uses for lasers. In dermatology, lasers are used to treat certain forms of skin cancer and to remove disfiguring marks such as "port wine" stains and tattoos. In neurosurgery and urology, lasers are used to remove certain tumors in otherwise inaccessible regions of the body and to break up kidney stones (lithotripsy). In gastroenterology, lasers are used to control bleeding from ulcers. In cardiovascular medicine, lasers can be used to remove obstructions in arteries (angioplasty). In oncology, lasers can optically activate cancerous cells previously tagged with hematoporphyrin dye, leading to the formation of chemically active molecules that selectively kill the cancerous cells. Various other medical applications of lasers in the fields of pulmonology, urology, gynecology, and orthopedic surgery are in early stages of development and approval. The application of lasers to medicine and dentistry continues to grow as the physiological effects of laser radiation become better understood and as equipment is developed to allow surgeons increased control over position, energy level, wavelength, and duration of tissue irradiation.

The manufacture of medical lasers and systems is dominated by ophthalmic applications, accounting for about $70 million of the total annual sales of about $220 million in 1988. Given the broadly successful application of lasers to surgical procedures, it is noteworthy that medical laser systems cannot yet be viewed a business success (a situation whimsically described as "profitless prosperity"). The lack of profitability stems from several causes, including the enormous effort required to bring a laser product to the medical marketplace (which requires FDA approval) and to gain acceptance by surgeons. Other causes are the competition from rapidly advancing, alternative technologies and the large number of competitors willing to enter an oversubscribed market. Despite the intense competition, the huge potential market for medical lasers will likely continue to attract manufacturers of laser systems. Of the roughly 10 million surgeries performed annually in the United States, fewer than 1% presently involve lasers. Advocates of laser-mediated surgery argue that as many as half of all surgeries might eventually be performed by or with the help of lasers. Because laser procedures are often less invasive than alternatives and can lead to rapid recovery with minimal or no hospital care, laser-based surgery may greatly expand as medical compensation paid by government and private insurance moves to a fixed allowance.

The Future

As numerous and diverse as current uses for the laser are, they appear to be only the advance guard of myriad applications now appearing on the horizon. Applications—both predicted and yet-unforeseen will multiply as researchers refine and enhance the laser's characteristics. We can expect future lasers to have increased efficiency, wavelengths extending from the ultraviolet to x-rays and even gamma rays, improved reliability and compactness, and higher powers. Lasers will become more versatile as they become more portable and as technology for projecting laser energy even more precisely becomes available. One important anticipated area of application is in lithography, where feature sizes below 0.1 micrometer could become practical. This technology could have important applications in future semiconductor chip manufacture. Their utility will expand as we increase their efficiency by coupling their energies to targets or material more effectively, for example through resonant or selective absorption.

The cost of lasers will decline as mass manufacturing leads to interchangeable laser modules (standardized as to dimensions and characteristics). The modular approach is of particular importance to semiconductor devices, because it will reduce cost and the need for tolerance control, but there are similar advantages to applying modular techniques to higher-power, high-energy devices for fusion and directed-energy weapons (for example, in allowing for graceful degradation while avoiding single-point failures).

As lasers themselves improve, advances in other technologies will create opportunities for powerful new combinations. For instance, fiber optics will be used with lasers not only to transmit information (as in communications or high-definition television) but also in sensors. Such systems will also increasingly replace electrical connections as a vehicle for transmitting energy. Integrated optics, a hybrid technology in which lasers are incorporated into semiconductor chips, will have many applications including optical computing and signal analysis. Optoelectronic connections in computers promise improved bandwidth (allowing higher speed), reduced size and weight, reduced susceptibility to electromagnetic interference, and reduced thermal dissipation (because of the elimination of hard wires and contacts). Nonlinear optics is another adjunct technology now being aggressively developed that will combine with the laser to offer novel applications in science, medicine, education, consumer goods, and commercial endeavors. Advances in materials science will allow the creation of new materials with specific chemical, electrical, and physical properties that will be beneficial to laser technology.

In many laser applications that require optical pumping, we expect that semiconductors lasers will increasingly be used as pump light sources. The increased demand for these lasers will in turn lead to greatly improved efficiency and reliability as well as reduced size. The semiconductor laser will likely replace other types of lasers in many other applications as well, as a result of two major developments. First, the emission from the laser diode can be shifted (by nonlinear conversion) from the infrared to the visible. Second, large and relatively inexpensive arrays of laser diodes emitting many watts of continuous-wave or pulsed power have been developed. The output of these arrays can be electronically steered by locking and electronically adjusting the phase between individual diode emitters. As a result, compact, low-cost, high-power laser sources emitting in the red, green, and blue spectral ranges will become available.

Applications of lasers can be foreseen in many areas: safety, where they can be used for windshear analysis, collision avoidance, or as simple sensors; security in low-light-areas; and forensics, where laser-induced fluorescence recovery of fingerprints is possible. Further advances in computer technology for image processing portend a major impact of lasers in military surveillance. The ready availability of less costly, shorter-wavelength sources will improve the technology of areas such as lithography, as noted above.

Certainly the use of laser-optic devices in adverse environments will increase; their resistance to electromagnetic interference, their nonelectrical contacts, and their ease of remote access are attractive advantages. These are applications in which the laser has a unique role.

As our ability to tune the wavelengths of high-power lasers increases, lasers will become more useful in the selective excitation of chemical reactions, and particularly in enhancing material deposition processes for thin films—for example, in superconducting films. Frequency tunability will also figure in a major new area: nonresonant and resonant interaction in various systems, as in the control of gas discharges through optogalvanic processes.

In medicine and biology, lasers with more efficient, shorter wavelengths (to x-rays and beyond) will make it possible to examine the structure of biological entities and (coupled with short-pulse capabilities) to analyze the kinetics of structures. Furthermore it will be feasible to modify such structures and explore them through cineradiography. Lasers will also be used in biological research for the selective destruction, modification, sorting and counting of biologically interesting species.

Clearly we have only begun to feel the impact the laser will have on our lives and civilization.

Bibliography

Jesse H. Ausubel and H. Dale Langford, eds., *Lasers, Invention to Application* (Washington, D.C.: National Academy Press, 1987).

Morris I. Cohen, ed., *Proc. IEEE* 70 (June 1982), special issue on laser applications.

John N. Howard, ed., *Optics Today*, readings from *Physics Today* (New York: American Institute of Physics, 1986).

Henry Kressel, ed., *Semiconductor Devices for Optical Communications* (New York: Springer-Verlag, 1980).

Journal of Quantum Electronics QE-20 (June 1984), IEEE Centennial Issue.

Physics Today (May 1976), special issue on lightwave communications.

Physics Today (May 1977), special issue on applications of lasers in research.

Physics Today (May 1985), special issue on optoelectronics.

Notes

In these notes, references to "paraphrases" of interviews in the Sources for the History of Lasers (SHL) documents indicate that the person interviewed has read and revised the transcript and that the quotes are treated here as quotes from written memoirs.

Preface

1. Paul Forman, "The First Atomic Clock Program: NBS, 1947–1954," *Proc. 17th Annual Precise Time & Time Interval (PTTI) Applications and Planning Meeting*, Washington, D.C., 3–6 December 1985; "Atomichron®: The Atomic Clock from Concept to Commercial Product," *Proc. IEEE* 73 (July 1985), 1181–1204; working files for the exhibits "Atomic Clocks" and "Science, Power, and Conflict," Division of Electricity and Modern Physics, National Museum of American History, Smithsonian Institution.

2. *Historical Studies in the Physical and Biological Sciences* 18(1) (1987), 149–229.

3. *Historical Studies in the Physical and Biological Sciences* 18(1) (1987), 111–147; *Physics Today* 41 (October 1988), 12–19.

Chapter 1: Introduction

1. Edwin Mansfield, *The Economics of Technological Change* (New York: Norton, 1968), p. 54. These numbers are in current dollars.

2. A. G. Kenwood and A. L. Lougheed, *The Growth of the International Economy, 1820–1980* (London: George Allen and Unwin, 1983), p. 274.

3. EIA, *Electronics Industries Yearbook*, 1968, p. 2, converted to constant dollars with *U.S. Statistical Abstracts*, 1987, table 763, p. 454.

4. Stephen E. Ambrose, *Rise to Globalism* (London: Penguin Press, 1971), pp. 210, 222.

5. Ibid.

6. Lawrence Lessing, "The Electronics Era," *Fortune* 44 (July 1951), 78–83.

7. For the expansion of electronics in the 1950s and the military's role, see Harold G. Vatter, *The U.S. Economy in the 1950s: An Economic History* (New York: Norton, 1963).

8. Mansfield, *Economics*, cited in note 1, p. 163. The figures for 1948 and 1956 are not entirely comparable because the numbers include "pay and allowances of military R&D personnel beginning in 1953 and support from procurement appropriations of development, test, and evaluation beginning in 1954."

9. See the first section of chapter 2.

10. "ONR Maser Contracts" and "Air Force Contractors (and suggestions)," in file "Conference on Quantum Electronics (Steering Committee, 1959)," C. H. Townes Papers; see also the first section of chapter 2 in this book.

11. The 1950 figure is from *Industrial Research Laboratories of the United States*, 9th edition, 1950, Bulletin of the National Research Council, No. 120, November 1950, which has 2,895 entries. The 1960 figure is from Mansfield, *Economics*, cited in note 1, p. 45.

12. John E. Tilton discusses the military market and venture capital in the formation of new 1950s firms in the semiconductor industry in *International Diffusion of Technology* (Washington, D.C.: Brookings Institution, 1971).

13. Herman O. Steckler, *The Structure and Performance of the Aerospace Industry* (Berkeley: University of California Press, 1965).

14. R. R. Nelson and R. N. Langlois, "Industrial Innovation Policy: Lessons from American History," *Science* 219 (18 February 1983), 814–818.

15. Mansfield, *Economics*, cited in note 1, p. 54; Marsh W. White, "American Physicists in the Current Quarter Century," *Physics Today* 9 (January 1956), 32–36.

16. My discussion of climate is indebted to the work of Margaret B. W. Graham: "Industrial Research in the Age of Big Science," in Richard S. Rosenbloom, ed., *Research on Technological Innovation, Management and Policy*, vol. 2 (Greenwich, Connecticut: Jai Press, 1985), 47–79; "Corporate Research and Development: The Latest Transformation," *Technology and Society* 7 (1985), 179–195; and *RCA and the Video Disc* (New York: Cambridge University Press, 1986).

17. *New York Times Index* 1968, "Research," p. 1240. Robert Sigethy, *The Air Force Organization for Basic Research, 1945–1970* (Ann Arbor: University Microfilms, 1980), pp. 63–64.

18. Paul B. Stares, *The Militarization of Space: U.S. Policy 1945–1984* (Ithaca: Cornell University Press, 1985).

19. Robert E. Osgood, "The Military Issues," in Robert E. Osgood et al., *America and the World: From the Truman Doctrine to Vietnam* (Baltimore: Johns Hopkins, 1970), pp. 189–235.

20. *Electronics*, passim, 1963–1966.

21. *Electronics* 39 (21 February 1966), 23, 38, 40; 39 (7 March 1966), 40–41; 39 (4 April 1966), 151ff.

22. *Electronics Industry Yearbook*, 1971 and 1972; National Science Foundation, "National Patterns of R&D Resources, Funds, & Personnel in the United States, 1953–1978/79," NSF 78–313.

23. Francis E. Rourke, "The Domestic Scene," in Osgood et al., *America and the World*, cited in note 19, pp. 147–188.

24. National Science Foundation, "National Patterns of R&D Resources," cited in note 22.

25. *Laser Focus* 7 (January 1971), 24–25; 8 (January 1972), 24. It is not clear from these articles exactly what the editors include, but these figures appear to represent hardware sales.

26. The implication for policy is that a still more diverse set of research institutions might foster a still greater multiplicity of research approaches. One could envision, for example, laboratories funded by consumers.

27. Figures published in the trade press all are in this range, although the actual numbers vary widely.

Chapter 2: The Maser

1. Mario Bertolotti, *Masers and Lasers: An Historical Approach* (Bristol: Adam Hilger, 1983); see my review in *Technology & Culture* 26 (1985), 134–135.

2. Henry Guerlac, "The Radio Background of Radar," *Journal of the Franklin Institute* 250 (1950), 285–308; David Kite Allison, *New Eye for the Navy: The Origin of Radar at the Naval Research Laboratory* (Washington, D.C.: Naval Research Laboratory, 1981).

3. C. H. Townes, "Microwave Spectroscopy," *American Scientist* 40 (1952), 270–290, quotation on p. 273.

4. Walter Gordy, "Microwave Spectroscopy," *Review of Modern Physics* 20 (October 1948), 668–717, quotation on p. 668.

5. Richard Honerjaeger, "Mikrowellenspektroskopie," *Naturwissenschaften* 38(2) (1951), 34–39; A. H. Nethercot jr. et al., "Spectroscopy near the Boundary between the Microwave and Infrared Regions," *Nuovo Cimento* (series 9) 9 (1952), Supp. 3, 358–363.

6. O. R. Gilliam et al., "Microwave Spectroscopy in the Region from Two to Three Millimeters," *Physical Review* 78 (1950), 140–144; J. A. Klein et al., "Magnetron Harmonics at Millimeter Wavelengths," *Review of Scientific Instruments* 23 (February 1952), 78–82.

7. John R. Pierce, "Millimeter Waves," *Physics Today* 3 (November 1950), 24–29.

8. Minutes of the Meeting of the ONR Advisory Committee on Millimeter Waves, 25 April 1951, C. H. Townes Papers; Memo from John J. Slattery to Director of Engineering, Evans Signal Laboratory, 23 January 1950, Signal Corps Records, Federal Records Center, St. Louis, Missouri; Project Hartwell, MIT, "A Report on Security of Overseas Transport," 21 September 1950, MIT Lincoln Laboratory Library.

9. James B. Wiesner, "A Successful Experiment," *Naval Research Reviews* (July 1966); *Proceedings of the Fortieth Anniversary Symposium of the Joint Services Electronics Program (JSEP)*; David Robb and Arnold Shostak, eds., *The Joint Services Electronics Program, 40th Anniversary*, ADA 171 610, September 1986 (all three courtesy of Paul Forman); Daniel J. Kevles, *The Physicists* (New York: Knopf, 1978), pp. 352–356. Subsequent to Stanford, still other JSEP laboratories were added.

10. W. Henry Lambricht, *Governing Science and Technology* (New York: Oxford University Press, 1976), pp. 18–19. Allan Needell, "Preparing for the Space Age: University-Based Research, 1946–1957," *Historical Studies in the Physical and Biologi-*

cal Sciences 18(1) (1987), 89–109. For the Stanford Electronics Laboratory, see Stuart W. Leslie and Bruce Hevly, "Steeple Building at Stanford: Electrical Engineering, Physics, and Microwave Research," *Proc. IEEE* 73 (1985), 1169–1180; Stuart W. Leslie, "Playing the Education Game To Win: The Military and Interdisciplinary Research at Stanford," *Historical Studies in the Physical and Biological Sciences* 18(1) (1987), 55–88.

11. C. H. Townes Papers, files "US Navy" and "ONR Advisory Committee on Millimeter Wave Generation."

12. Townes's curriculum vitae is in the Sources for the History of Lasers. See also Theodore Berland, *The Scientific Life* (New York: Coward, 1962), chapter 3, "'Discovery Is a Revelation': Charles Hard Townes," pp. 70–101.

13. Townes's Eureka moment on a Washington park bench has been written up in many places. See, for example, the account by Townes's hotel roommate Arthur L. Schawlow in "Masers and Lasers," *IEEE Transactions on Electron Devices*, ED-23 (1976) 773–779; Berland, *The Scientific Life*, cited in note 12; and C. H. Townes, "Harnessing Light," *Science '84* 5 (November 1984), 153–155.

14. Pierce, "Millimeter Waves," cited in note 7. These events repeated the pattern of the 1930s when the magnetron was pursued because the "conventional" vacuum tube gave unsatisfactory performance for the centimeter wave region. See James E. Brittain, "The Magnetron and the Beginnings of the Microwave Age," *Physics Today* 38 (July 1985), 60–67. For the idea, mentioned by Brittain and also pertinent here, of "presumptive anomaly," see Edward Constant, *The Origins of the Turbojet Revolution* (Baltimore: Johns Hopkins, 1980), pp. 12–13.

15. Townes was to create this abnormal population distribution in his ammonia beam masers, as we shall see, by physically separating the molecules in the high energy state from those in the low energy state with a "focuser." It would eventually become more common to create inverted populations by raising the energy of an aggregate of molecular systems. The aggregate energy can be pumped up in a variety of ways, among them irradiation with light (optical pumping), transfer of energy to the lasing atoms or molecules from highly energetic electrons (electron excitation), and transfer of the energy of chemical bonds (chemical lasers).

16. "The Invention of the Maser and Laser" (n.d.), C. H. Townes Papers, File "Maser-laser history." C. H. Townes, interview by W. V. Smith, June 1979, Niels Bohr Library, American Institute of Physics.

17. Charles H. Townes, "Ideas and Stumbling Blocks in Quantum Electronics," *IEEE Journal of Quantum Electronics* QE-20 (1984), 548. The narrow frequency width of the maser was imposed by the resonant cavity, which selected a discrete set of all allowable frequency intervals out of the transition's total line width.

18. Ibid., p. 548.

19. The relevant part of their paper is reprinted in Willis E. Lamb jr., "Physical Concepts in the Development of the Maser and Laser," in Behram Kursunoglu and Arnold Perlmutter, eds., *Impact of Maser Research On Technology* (New York: Plenum, 1973), pp. 80–85.

20. E. M. Purcell and R. V. Pound, "A Nuclear Spin System at Negative Temperature," *Physical Review* 81 (1951), 279–280. See C. G. B. Garrett, *Gas Lasers* (New York: McGraw-Hill, 1967), p. vii.

21. Townes discusses this point in "Ideas and Stumbling Blocks," cited in note 17.

22. H. W. Schulz to J. N. Wickert, 14 August 1947, memorandum, "Catalysis by Electromagnetic Activation" (courtesy of A. L. Schawlow); file folder "Carbide and Carbon," C. H. Townes Papers.

23. James P. Gordon, interview by Paul Forman, 4 November 1983 (SHL).

24. Joseph Weber, interview, 8 April 1983 (SHL).

25. Weber, interview, cited in note 24; J. Weber to A. L. Schawlow, 8 April 1963, and J. Weber to C. H. Townes, 14 November 1958, in C. H. Townes Papers, files "Schawlow" and "Weber," respectively; J. Weber to J. L. Bromberg, n.d. (ca. April 1986) (SHL).

26. J. Weber, "Amplification of Microwave Radiation by Substances Not in Thermal Equilibrium," *Transactions of the IRE Professional Group on Electron Devices* PGED-3 (June 1953), 1–4.

27. Weber to Schawlow, cited in note 25.

28. See, for example, the interview with Lawrence Goldmuntz (SHL); Bertolotti, *Masers and Lasers*, cited in note 1, p. 75.

29. H. Friedburg and W. Paul, "Optische Abbildung mit neutralen Atomen," *Naturwissenschaften* 38 (1951), 159–160. To adapt the focuser for molecules, Townes had to convert it from a device using magnetic fields to one using electric fields.

30. Quarterly Report of the Columbia Radiation Laboratory, 31 December 1951, pp. 7–8 (Columbia University Archives). The calculations were made by George Dousmanis and Herbert Zeiger.

31. Quarterly Report of the Columbia Radiation Laboratory, 31 March 1952, p. ll; C. H. Townes, deposition in *Research Corporation* v. *Spectra-Physics*, 8 July 1970 (courtesy of Dana Raymond).

32. Gordon, interview, cited in note 23; C. H. Townes, "Computation Book III," C. H. Townes Papers; Townes, deposition cited in note 31; J. P. Gordon, "Hyperfine Structure in the Inversion Spectrum of $N^{14}H_3$ by a New High-Resolution Microwave Spectrometer," *Physical Review* 99 (1955), 1253–1263; J. P. Gordon, H. J. Zeiger, and C. H. Townes, "Molecular Microwave Oscillator and New Hyperfine Structure in the Microwave Spectrum of NH_3," *Physical Review* 95 (1954), 282–284, and "The Maser—New Type of Microwave Amplifier, Frequency Standard, and Spectrometer," *Physical Review* 99 (1955), 1264–1274. The quotation is from the Quarterly Report of the Columbia Radiation Laboratory, 30 April 1954, p. 8.

33. Paul Forman, "Atomichron®: The Atomic Clock from Concept to Commercial Product," *Proc. IEEE* 73 (1985), 1181–1204.

34. Martin Packard and Russell Varian, "Free Nuclear Induction in the Earth's Magnetic Field (Abstract)," *Physical Review* 93 (1954), 941.

35. Forman, "Atomichron®," cited in note 33.

36. E. C. Bullard, "Definition of the Second of Time," *Nature* 176 (13 August 1955), 282; C. T. Clark et al., "Navaglobe-Navarho Long-Range Radio Navigational System," *Electrical Communication* 31 (1954), 155–166; E. A. Gerber, "Precision Frequency Controls for Guided Missiles," 29 July 1957, U.S. Army Signal Engineer-

ing Laboratories, Fort Monmouth, New Jersey. (I am grateful to Paul Forman for the last two items.)

37. Walter H. Higa, "A History of Masers at the Jet Propulsion Laboratory," 18 March 1988 (SHL); Quarterly Report of the Columbia Radiation Laboratory, 30 October 1954, p. 14.

38. Willis E. Lamb jr., "Laser Theory and Doppler Effects," *IEEE Journal of Quantum Electronics* QE-20 (June 1984), 551–555.

39. John C. Helmer to J. L. Bromberg, 30 June 1986 and 5 August 1986 (SHL); Quarterly Report of the Columbia Radiation Laboratory, 28 February 1955, p. 19.

40. Charles H. Townes to Harold Lyons, 9 April 1954 (courtesy of H. Lyons).

41. Paul Forman, "The First Atomic Clock Program: NBS, 1947–1954," *Proc. 17th Annual Precise Time & Time Interval (PTTI) Applications and Planning Meeting*, Washington, D.C., 3–6 December 1985.

42. Manuscript histories, on file at the Archives of Hughes Aircraft Company; B. W. Henry (Hughes archivist) to J. L. Bromberg, private communication.

43. Research Laboratories, Annual Status Report for 1956 (courtesy of George F. Smith); Philip Klass, "Hughes Aircraft Co. Accents the New," *Aviation Week* 58 (29 June 1953), 49.

44. Harold Lyons to C. H. Townes, 8 May 1955; Lyons to Lt. Col. John V. Fill, and Lyons to E. A. Gerber, 9 May 1955; Hughes Research Laboratory Memorandum, "Informal Proposal" to Signal Corps Engineering Laboratories, 11 October 1955, H. Lyons papers (all courtesy of Paul Forman).

45. Winfield E. Fromm, "AIL in World War II," *AIL Record* (September 1970) (courtesy of Karle S. Packard); M. M. Freundlich et al., "Final Report on Molecular-Beam Frequency Standard," January 1956, extract (courtesy of Paul Forman); C. H. Townes to Sam Johnston, 1 April 1955; Martin M. Freundlich to C. H. Townes, 1 June 1955, and R. F. Simons to C. H. Townes, 23 November 1955, C. H. Townes Papers.

46. M. D. Fagen, ed., *A History of Engineering and Science in the Bell System: National Service in War and Peace (1925–1975)* (Bell Telephone Laboratories, 1978), p. 356.

47. *The New York Times*, 15 December 1955, p. 59.

48. M. J. Kelly, "Communications and Electronics," *Electrical Engineering* 71 (November 1952), 965–969, esp. 969. (Mervin J. Kelly was Bell Laboratories Director of Research.)

49. S. Millman, ed., *A History of Engineering and Science in the Bell System: Communications Sciences (1925–1980)* (AT&T Bell Laboratories, 1984), p. 264.

50. James P. Gordon, addendum (27 October 1986) to the interview for the Laser History Project (SHL).

51. J. P. Gordon to C. H. Townes, n.d. (probably fall 1955), C. H. Townes papers; Rudolf Kompfner, *The Invention of The Traveling-Wave Tube* (San Francisco: San Francisco Press, 1964).

52. Optical pumping was pioneered by Alfred Kastler and his associates, but Dicke and his graduate students also carried out some early independent work. See R. H.

Dicke, "Early Work at Princeton on Optical Pumping," in *Polarisation, Matière et Rayonnement* (Paris: Presses Universitaires de France, 1969), pp. 431–435.

53. No spectral line is infinitely narrow (monochromatic). But over and above the natural line breadth, there are a number of external effects that tend to broaden spectral lines, and these can be reduced; see C. H. Townes and A. L. Schawlow, *Microwave Spectroscopy* (New York: McGraw-Hill, 1955), pp. 336–375.

54. H. Y. Carr and E. M. Purcell, "Interaction between Nuclear Spins in HD Gas," *Physical Review* 88 (1952), 415–416.

55. R. H. Dicke, private communication; R. H. Romer and R. H. Dicke, "New Techniques for High-Resolution Microwave Spectroscopy," *Physical Review* 99 (1955), 532–536; R. H. Dicke and R. H. Romer, "Pulse Techniques in Microwave Spectroscopy," *Review of Scientific Instruments* 26 (1955), 915–928.

56. R. H. Dicke, "A Scientific Autobiography," January 1975, unpublished memoir, and correspondence, 1949–1954, passim, R. H. Dicke papers; Robert H. Dicke, interview by Paul Forman and Joan Bromberg, 2 May 1983 (SHL).

57. Anthony E. Siegman, *Lasers* (Mill Valley, California: University Science Books, 1986), p. 548; R. H. Dicke, "Coherence in Spontaneous Radiation Processes," *Physical Review* 93 (1954), 99–110.

58. Dicke's superradiance paper was submitted in August 1953, while he first learned of the maser in November 1953 when he gave a lecture at Columbia University (see the file "correspondence 1953" in R. H. Dicke papers and the interview with Dicke cited in note 57).

59. Paul Forman, "Behind Quantum Electronics: National Security as Basis for Physical Research in the United States, 1940–1960," *Historical Studies in the Physical and Biological Sciences* 18(1) (1987), 149–229; Dicke, "Idea Notebook," August 25, 1945, and forward; Dicke to Director, Squier Signal Laboratory, Fort Monmouth, December 27, 1950; and E. W. Engstrom to Dicke, 26 September 1955, R. H. Dicke Papers (the quotation is drawn from this letter).

60. L. E. Norton, "Coherent Spontaneous Microwave Emission by Pulsed Resonance Excitation," *IRE Transactions on Microwave Theory and Techniques* MTT-5 (1957), 262–265; Robert H. Dicke to Harold Lyons, 15 April 1955, and E. A. Gerber to Harold Lyons, 26 April 1955, H. Lyons Papers (courtesy of Paul Forman).

61. James P. Wittke, interview for the Laser History Project, 13 September 1983 (SHL); James P. Wittke, "Molecular Amplification and Generation of Microwaves," *Proc. IRE* 45 (1957), 291–316 (see pp. 307–308).

62. The United States was, of course, not the only focal point of maser research. Similar ideas were also being pursued in Moscow, at the Lebedev Physical Institute of the USSR Academy of Sciences. There, a group around Alexandr M. Prokhorov and Nikolai G. Basov, with an interest in more sensitive and higher-resolution spectrometers, had been led first to work on molecular beam spectrometers, then to a concentration on emission rather than absorption spectra, and finally to the idea of a stimulated-emission oscillator. By 1953, the Lebedev group had formed ideas on how to make a stimulated-emission oscillator, although they did not succeed in constructing one until after they had had an opportunity to read the Gordon-Zeiger-Townes paper. See N. G. Basov, "On the History of Lasers at P. N. Lebedev Physical Institute," an interview by Arthur H. Guenther for the Laser

History Project, 14 September 1984, published in *Kvantoya Elektronika* 12(3) (1985), 453–464 [trans.: *Soviet Journal of Quantum Electronics* 15(3) (1985), 301–307], and *Uspekhi Fiz. Nauk* 148(2) (1986), 313–324 [trans.: *Soviet Physics. Uspekhi* 29(2) (1986), 179–185]; Bertolotti, *Masers and Lasers*, cited in note 1, pp. 84–86.

63. Forman, "The First Atomic Clock Program," cited in note 41.

64. The contract was DA 36-039-SC-71178.

65. This paragraph is based on materials Paul Forman collected from Signal Corps records at the St. Louis Federal Records Center, particularly E. A. Gerber to V. Hughes, 13 August 1956; Memorandum Report on Contract DA 36-039-SC-73041; F. H. Reder to Chief, Frequency Control Branch Components Department, 13 July 1956; and "Suggested Sources for PR&C 56-ELS/R-3508," 1 March 1956.

66. Gordon, Zeiger, and Townes, "The Maser," cited in note 32, pp. 1272–1274.

67. Attempts were being made to run molecular beam masers with other gases, or on ammonia lines other than the 3-3 line, and these, of course, would have yielded other frequencies.

68. Wittke, "Molecular Amplification," cited in note 61, p. 308.

69. M. W. P. Strandberg, "Quantum Mechanical Amplifiers," *Proc IRE* 45 (1957), 92–93. The translation from power to bandwidth comes about because one can cause ions in different regions of the solid to have different frequencies by applying inhomogeneous magnetic fields and still have enough power to get adequate amplification.

70. C. H. Townes, private communication.

71. C. H. Townes, Computation Book III, entries of 22–29 December 1955, pp. 76–84. Townes to J. L. Bromberg, July 31, 1987, C. H. Townes Papers. Townes had briefly touched on spin masers in entries of 22 March 1954 and 1 July 1954, but it is characteristic of the interaction among scientists of different specialties in the maser/laser field that it took Honig's communication to Townes of the properties of a specific solid to raise this project from a notebook speculation to an experimental program.

72. Jean Combrisson, Arnold Honig, and Charles H. Townes, "Utilisation de la résonance de spins électroniques pour réaliser un oscillateur ou un amplificateur en hyperfréquencies," *Comptes Rendus de l'Académie des Sciences* (Paris) 242 (14 May 1956), 2451–2453.

73. A. L. Schawlow, deposition, 24 June 1963, Interference #92,015; Gordon, addendum to interview, cited in note 50; C. H. Townes to J. L. Bromberg, 8 May 1985, C. H. Townes Papers.

74. On the Research Laboratory of Electronics, see Forman, "Atomichron®," cited in note 33, and the references therein.

75. Strandberg, "Quantum Mechanical Amplifiers," cited in note 69.

76. M. W. P. Strandberg, interview, 15 June 1983 (SHL); MIT Research Laboratory of Electronics, Quarterly Progress Reports, 1955–1957, passim; M. W. P. Strandberg, computation books #1170 and #1121 (courtesy of W. P. Strandberg).

77. N. Bloembergen, Laboratory Notebook, 22 June 1956 and forward (courtesy of N. Bloembergen); N. Bloembergen, interview, 27 June 1983 (SHL); Benjamin Lax, interview, 15 May 1986 (SHL).

78. Three-level systems were also being suggested by others at this time both in notebooks (see, for example, the interview with Ali Javan in *Lasers and Applications: Laser Pioneer Interviews* [Torrance, California: High Tech Publications, 1985], p. 116) and in publications (N. G. Basov and A. M. Prokhorov, *Soviet Physics JETP* I (1955), 184–185). But these were gas systems and, even more to the point, were unknown to Bloembergen when he did his work.

79. Bertolotti, *Masers and Lasers*, cited in note 1, pp. 32–59.

80. *Physical Review* 73 (1948), 679–712.

81. Albert W. Overhauser, "Polarization of Nuclei in Metals," *Physical Review* 92 (1953), 411–415. Bertolotti, *Masers and Lasers*, cited in note 1, pp. 51–55.

82. Although a molecular system may exist in many energy states, from a given state, not every other state is accessible.

83. Bloembergen, notebook, cited in note 77. N. Bloembergen, "Proposal for A New Type Solid State Maser," *Physical Review* 104 (1956), 324–327.

84. R. Kompfner to Neville [Robertson], 19 July 1956, and R. Kompfner to "Sir," 27 July 1956, Rudolf Kompfner Papers, Box 8, AT&T Archives; G. Feher, J. P. Gordon, E. Buehler, E. A. Gere, and C. E. Thurmond, "Spontaneous Emission of Radiation from an Electron Spin System," *Physical Review* 109 (1958), 221–222.

85. H. E. D. Scovil, interview, 5 July 1983 (SHL).

86. H. E. D. Scovil, "A Continuously Operative Solid State Microwave Amplifier by Stimulated Emission of Radiation," Case 38728, 7 August 1956 (courtesy of H. E. D. Scovil).

87. Rudolf Kompfner, "Memorandum for File," 18 August 1956, R. C. Fletcher and R. Kompfner to H. Friis and J. A. Morton, 11 September 1956, and R. Kompfner, "The Solid State Laser. Memorandum for Record," 14 September 1956, Box 8, R. Kompfner Papers, AT&T Archives; Bloembergen, notebook, cited in note 77; interview with Scovil, cited in note 85.

88. Cruft Laboratory, Harvard University, Progress Reports No. 41 (1 July 1956–1 October 1956), p. 16; 42 (1 October 1956–1 January 1957), p. 19; and 43 (1 January 1957–1 April 1957), p. 19.

89. H. E. D. Scovil, G. Feher, and H. Seidel, "Operation of a Solid State Maser," *Physical Review* 105 (1957), 762–763.

90. Paraphrases of interviews with James W. Meyer, 20 November 1986, and Alan L. McWhorter, 1 April 1985 (SHL); Quarterly Progress Reports of the Solid State Research Group, November 1956 and forward (MIT Lincoln Laboratory Library Archives).

91. Its origins are described in the next section of this chapter.

92. Albert W. Schrader Papers, Michigan Historical Collections, Bentley Historical Library, University of Michigan, passim. The laboratory took its name from the Willow Run Airport, which the university purchased from the government in 1949 for $1 as surplus war equipment. The air field had been built during the war as a proving ground for the Ford-Built B-64 bomber.

93. Weston E. Vivian, paraphrase of an interview for the Laser History Project, 14 October 1986 (SHL). A "Statement of Work: Electron-Spin Resonance Amplifi-

ers," which internal evidence suggests dates from December 1956 or January 1957, specifies May 1956 for the inauguration of the program. The "University of Michigan Research Institute News" of December 1958 (IX #6) gives a date of early summer. (Both items courtesy of Chihiro Kikuchi.)

94. For a while, Kikuchi underlined the complementarity by renaming his first program MASAR, Microwave Attenuation by Stimulated Absorption of Radiation; "Proposal for the Investigation of Solid State Masar-Maser Systems" (probably late 1956 or early 1957; courtesy of C. Kikuchi).

95. Cruft Laboratory, Harvard University, Progress Report No. 42, 1 October 1956–1 January 1957, p. 19.

96. Gadolinium ethyl sulfate was widely recognized, at BTL and elsewhere, to have a number of unfavorable properties, such as mechanical softness.

97. Chihiro Kikuchi, paraphrase of an interview, 19 October 1985 (SHL).

98. Chihiro Kikuchi, "Ruby Maser and Laser," talk to the Ann Arbor Optimist Club, 9 October 1962 (courtesy of C. Kikuchi).

99. Chihiro Kikuchi, "Masar and Maser," talk at Sylvania, 26 April 1957 (courtesy of C. Kikuchi).

100. The pump, however, would be at 24,000 megahertz, where new expertise would be required.

101. See Kikuchi, "Proposal for the Investigation," cited in note 94.

102. J. W. Orton, D. H. Paxman, and J. C. Walling, *The Solid State Maser* (Oxford: Pergamon Press, 1970), pp. 37–42, 192–193; papers by G. Feher and H. E. D. Scovil, "Electron Spin Relaxation Times in Gadolinium Ethyl Sulfate," and by Scovil, Feher, and H. Seidel, "Operation of a Solid State Maser," are reprinted on pp. 194–203 (originally in *Physical Review* 105 [1957], 760–763); interview with Scovil, cited in note 85.

103. Alan L. McWhorter and James W. Meyer, "Solid-State Maser Amplifier," *Physical Review* 109 (1958), 312–318, and A. L. McWhorter, J. W. Meyer, and P. D. Strum, "Noise Measurement on a Solid-State Maser," *Physical Review* 108 (1957), 1642–1644, both reprinted in Orton, Paxman, and Walling, *The Solid State Maser*, cited in note 102, pp. 206–229, with commentary on pp. 204–205; J. W. Meyer, "The Historical Development of a Solid State Maser Amplifier," Group Report M37-17, 19 July 1957, MIT Lincoln Laboratory (courtesy of J. W. Meyer); James W. Meyer, paraphrase of an interview, 20 November 1986 (SHL).

104. C. H. Townes to A. Kastler, 10 August 1957, C. H. Townes Papers; J. O. Artman et al., "Operation of a Three-Level Solid-State Maser at 21 cm," *Physical Review* 109 (1958), 1392–1393; *Science* 127 (10 January 1958), 77.

105. G. Makhov, C. Kikuchi, J. Lambe, and R. W. Terhune, "Maser Action in Ruby," *Physical Review* 109 (1958), 1399–1400.

106. A number of other groups had been considering ruby during 1957, at laboratories including Harvard, Stanford, and Bell, but none, to my knowledge, made it the focus of their maser work.

107. J. R. Pierce to R. Q. Twiss, 19 October 1956, R. Kompfner Papers, Box 8, AT&T Archives.

108. Heffner to Bloembergen, 4 October 1956, N. Bloembergen Papers, Harvard University.

109. A. E. Siegman, interview, 19 August 1986, untranscribed tape, Niels Bohr Library, American Institute of Physics.

110. Stanford Electronics Laboratories, Electron Tube Laboratory, "Electron Tube Research, Consolidated Quarterly Status Report," nos. 1–5 (March 1957–March 1958).

111. A. Michal McMahon, *The Making of A Profession: A Century of Electrical Engineering in America* (New York: IEEE Press, 1984); *Proc. IEEE* 59 (June 1971), 820–1016 (special issue on engineering education); Carl Barus, "Military Influence on the Electrical Engineering Curriculum since World War II," *IEEE Technology and Society Magazine* 6 (June 1987), 3–8.

112. This is the treatment given by Donald O. Pederson, Jack J. Studer, and John R. Whinnery in *Introduction to Electronic Systems, Circuits, and Devices* (New York: McGraw-Hill, 1966), a textbook for a first course at the sophomore or junior level whose principles were formulated by the Electrical Engineering Department at the University of California, Berkeley.

113. A. E. Siegman, interview, 23 January 1984 (SHL).

114. Robert W. Hellwarth, interview, 28–29 May 1985 (SHL).

115. Amnon Yariv, interview, 28 January 1985 (SHL).

116. Joseph E. Geusic, paraphrase of an interview, 21 October 1987 (SHL).

117. R. H. Pantell, private communication, 23 January 1984.

118. All of this paragraph and the next are based upon Anthony E. Siegman, "Thermal Noise in Microwave Systems," *Microwave Journal* 4 (1961), part I: March, 81–90, part II: April, 66–73, part III: May, 93–104.

119. Cited by Siegman as H. A. Haus and R. B. Adler, "An Extension of the Noise Figure Definition," *Proc. IRE* 45 (1957), 690–691, and "Optimum Noise Performance of Linear Amplifiers," *Proc. IRE* 46 (1958), 1517–1533.

120. Robert Serber and C. H. Townes, "Limits on Electromagnetic Amplification due to Complementarity," in Charles H. Townes, ed., *Quantum Electronics* (New York: Columbia University Press, 1960), p. 233; H. Friedburg, "General Amplifier Noise Limit," ibid., p. 228.

121. H. A. Haus and J. A. Mullen, "Equivalent Circuits for Quantum Noise in Linear Amplifiers," in P. Grivet and N. Bloembergen, eds., *Quantum Electronics: Proceedings of the Third International Congress, Paris* (New York and Paris: Columbia University Press and Dunod, 1964), vol. I, pp. 71–93.

122. E. I. Gordon, "Optical Maser Oscillators and Noise," *Bell System Technical Journal* 43 (1964), 507.

123. See P. N. Butcher, "An Introduction to the Theory of Solid-State Masers, . . . ," *Proc. IEE* 107B (1960), 341–351, especially p. 347.

124. H. Motz, "Negative Temperature Reservoir Amplifiers," *Journal of Electronics* 2 (1957), 571–578; R. Kompfner, "Memorandum for Record, Traveling-Wave Type of S.S. Maser with Gain in One Direction Only," 16 October 1956, Case 38543, R. Kompfner Papers, Box 8, AT&T Archives.

125. W. S. C. Chang, J. Cromack, and A. E. Siegman, "Cavity Maser Experiments Using Ruby at S Band," *Journal of Electronics and Control* 6 (1959), 508–526.

126. McWhorter, paraphrase of an interview, cited in note 90.

127. R. W. DeGrasse, E. O. Schulz-DuBois, and H. E. D. Scovil, "The Three-Level Solid State Traveling-Wave Maser," *Bell System Technical Journal* 38 (1950), 305–334, reprinted in Orton, Paxman, and Walling, *The Solid State Maser*, cited in note 102, pp. 234–268; H. Seidel, private communication; Scovil, interview, cited in note 85.

128. H. I. Ewen and E. M. Purcell, "Observation of a Line in the Galactic Radio Spectrum.: Radiation from Galactic Hydrogen at 1,420 Mc/s" *Nature* 168 (1 September 1951), 356; F. T. Haddock and T. P. McCullough jr., "Extension of Radio Source Spectra to a Wavelength of 3 Centimeters," *Astronomical Journal* 60 (1955), 161–162.

129. J. A. Giordmaine, "Centimeter Wavelength Radio Astronomy Including Observations Using the Maser," *Proceedings of the National Academy of Sciences (USA)* 46 (1960), 267–276.

130. Peter D. Strum, "Considerations in High-Sensitivity Microwave Radiometry," *Proc. IRE* 46 (January 1958), 43–53.

131. Airborne Instruments Laboratory, "Circulator Maser System," *Proc. IRE* 46 (June 1959), 4A; R. Kompfner to H. T. Friis, 13 March 1957, R. Kompfner papers, AT&T Archives.

132. H. E. D. Scovil, "On the Application of Microwave and Optical Masers," in R. E. Burgess, ed., *Progress in Radio Science 1960–1963*, Vol. VII, *Radio Electronics* (Amsterdam: Elsevier, 1965), pp. 129–157.

133. Bloembergen, notebook, cited in note 77; N. Bloembergen to P. W. Levy, 27 December 1956, from "Chronological files, 1952–1963," N. Bloembergen Papers, Harvard University.

134. Joseph A. Giordmaine, interview, 31 May 1984 (SHL); Giordmaine, "Centimeter Wavelength Radio Astronomy," cited in note 129. See also Cornell H. Mayer, paraphrase of an interview, 8 November 1986 (SHL).

135. This was common for radioastronomers. See David O. Edge and Michael J. Mulkay, *Astronomy Transformed: The Emergence of Radio Astronomy in Britain* (New York: Wiley, 1976).

136. F. T. Haddock, private communication, 14 October 1986.

137. "University of Michigan Research Institute News," cited in note 93; "Project MICHIGAN Maser Research and Development," n.d. (probably 1958; courtesy of C. Kikuchi).

138. J. J. Cook et al., "Radio Detection of the Planet Saturn," *Nature* 188 (29 October 1960), 393–394; William E. Howard III, Alan H. Barrett, and Fred T. Haddock, "Measurement of Microwave Radiation from the Planet Mercury, *Astrophysics Journal* 136 (1962), 995–1004.

139. G. Wade, "Low-Noise Amplifiers for Centimeter and Shorter Wavelengths," *Proc. IRE* 49 (1961), 880–891.

140. W. H. Higa, "Theory of a New Low Noise Amplifier," October 1956, typescript (SHL).

141. According to W. H. Louisell and C. F. Quate, "Parametric Amplification of Space Charge Waves," *Proc. IRE* 46 (1958), 707.

142. H. Suhl, "Proposal for a Ferromagnetic Amplifier in the Microwave Range," *Physical Review* 106 (1957), 384–385, and "Theory of the Ferromagnetic Microwave Amplifier," *Journal of Applied Physics* 28 (1957), 1225–1236. Electron tube noise comes from the "shot effect" due to the discrete nature of electrons, and from the fact that electrons have to be emitted by a hot cathode, so that it is impossible to reduce the thermal noise by cooling the apparatus. See H. Heffner and G. Wade, "Gain, Bandwidth, and Noise Characteristics of the Variable-Parameter Amplifier," *Journal of Applied Physics* 29 (1958), 1321–1331.

143. Max T. Weiss, "A Solid-State Microwave Amplifier and Oscillator Using Ferrites," *Physical Review* 107 (1957), 317.

144. *Aviation Week and Space Technology* 69 (14 July 1958), 83; (4 August 1958), 69ff.; (18 August 1958), 81; (1 September 1958), 64–65; (8 September 1958), 64ff.

145. Wade, "Low-Noise Amplifiers," cited in note 139.

146. Giordmaine, interview, cited in note 134.

147. Haddock, private communication, cited in note 136.

148. James W. Meyer, "Systems Applications of Solid-State Masers," *Electronics* 33 (4 November 1960), 58–63.

149. B. J. Robinson, "Low-Noise Amplifiers in Radio Astronomy," *Progress in Radio Science, 1963–1966, Proc. XVth General Assembly of URSI, Munich, 5–15 September 1966,* pp. 2062–2089, especially pp. 2067–2075.

150. H. R. Senf et al., "Masers for Radar Systems Applications," *IRE Transactions on Military Electronics* MIL-5 (1961), 58–65.

151. Philip J. Klass, "New Technology Boosts Interceptor Role," *Aviation Week and Space Technology* 69 (10 November 1958), 31. The F-108 itself was canceled in 1962, with serious effects on Hughes's business.

152. Hughes Aircraft Company, "Research Laboratories' Annual Technical Status Report for 1957" and "Research Laboratories' Annual Status Report, 1 July 1957 through 30 June 1958" (courtesy of George F. Smith).

153. Theodore H. Maiman, interview by James Cavuoto, *Lasers & Applications* IV (May 1985), 85–90.

154. F. E. Goodwin, "Duplexing a Solid-State Ruby Maser in an X-Band Radar System," *Proc. IRE* 48 (1960) 113; J. L Carter, M. Katzman, and I. Reingold, "The Use of an X-Band Solid-State Ruby Maser with a Conventional Duplexing System," *Microwave Journal* 3 (July 1960), 43–46; H. R. Senf, F. E. Goodwin, and J. E. Kiefer, "Masers for Radar Systems Applications," *IRE Transactions on Military Electronics,* MIL-5 (1961), 58–65.

155. Massachusetts Institute of Technology, "Review Panel on Special Laboratories, Final Report," October 1969, MIT Humanities Library; William H. Radford, "M.I.T. Lincoln Laboratory: Its Origin and First Decade," *Technology Review* 64 (January 1962), 15–18. Funding for the laboratory came from the Air Force, Army, and Navy through an Air Force contract.

156. Benjamin Lax, interview, 15 May 1986 (SHL); MIT Lincoln Laboratory. Quarterly Progress Reports, Solid State Research, November 1956 through #4, 1962 (MIT Lincoln Laboratory Library Archives).

157. R. Price et al., "Radar Echoes from Venus," *Science* 129 (20 March 1959), 751–753; "Radar Echoes from Venus: Press Kit" (MIT Lincoln Laboratory Library Archives). It was later established that these returns were spurious.

158. Memorandum from G. H. Pettengill to J. V. Harrington, 8 July 1959, Venus shot records (MIT Lincoln Laboratory Library Archives).

159. Haddock, private communication, cited in note 136; Cornell H. Mayer, private communication, 8 November 1986. The way in which one scientific administrator, Lloyd Berkner, perceived radioastronomy as furthering, at one and the same time, science, the military, and national prestige is well laid out in Allan A. Needell, "Lloyd Berkner, Merle Tuve, and the Federal Role in Radio Astronomy," *Osiris* 3 (1987), 261–288.

160. J. R. Pierce, *The Beginnings of Satellite Communications* (San Francisco: San Francisco Press, 1968).

161. *New York Times*, 9 June 1955, p. 1.

162. J. R. Pierce, "Orbital Radio Relays," *Jet Propulsion* 25 (1955), 153–157, reprinted in *Beginnings of Satellite Communication*, cited in note 160.

163. Delbert D. Smith mentions Lockheed and RCA in his book *Communication via Satellite: A Vision in Retrospect* (Leyden and Boston: Nijhoff, 1976), p. 47.

164. J. R. Pierce and R. Kompfner, "Transoceanic Communication by Means of Satellites," *Proc. IRE* 47 (1959), 372.

165. Pierce, *Beginnings of Satellite Communication*, cited in note 160; "Report and Plans for Panel 4, (Electronics, Guidance and Control), of the NAS-ARDC 1958 Summer Study Group," dated 28 April 1958, and Theodore von Karman to R. Kompfner, 2 June 1958, R. Kompfner Papers, Box 8, AT&T Archives.

166. Pierce, *Beginnings of Satellite Communication*, cited in note 160; R. Kompfner to Director, Advanced Research Projects Agency, 24 July 1958, and Roy W. Johnson to R. Kompfner, 25 August 1958, R. Kompfner Papers, Box 6, AT&T Archives; Smith. *Communication via Satellite*, cited in note 163, pp. 51–54.

167. Millman, ed., *History*, cited in note 49, pp. 225–227; *Bell Systems Technical Journal* 40 (July 1961), 975–1238 (special issue on Project Echo); Scovil, interview, cited in note 85; D. C. Hogg and H. E. D. Scovil, "Measuring the Sky's Electrical Noise," *Bell Laboratories Record* 39 (August 1961), 276–279.

168. Jonathan F. Galloway, *The Politics and Technology of Satellite Communications* (Lexington, Massachusetts: Lexington Books, 1972), pp. 39–41.

169. Gerald Brock, *The Telecommunications Industry* (Cambridge: Harvard University Press, 1981).

170. Smith, *Communication via Satellite*, cited in note 163, pp. 58–60, 70–71.

171. L. W. Davies to E. G. Bowen, 19 February 1959; Kompfner to Bowen, 18 November 1959, R. Kompfner Papers, Box 6, AT&T Archives.

172. R. Kompfner to A. C. B. Lovell, 1 March 1961, R. Kompfner Papers, Box 5, AT&T Archives.

173. Jeremy Bernstein, *Three Degrees Above Zero: Bell Labs in the Information Age* (New York: Scribner, 1984), pp. 213–232; A. A. Penzias, private communication, 6 December 1988.

174. Millman, ed., *History*, cited in note 49, p. 228; Pierce, *Beginnings of Satellite Communication*, cited in note 160, p. 28; Smith, *Communication Via Satellite*, cited in note 163, and Galloway, *Politics and Technology*, cited in note 168, passim.

175. R. W. Wilson, "The Cosmic Microwave Background Radiation," *Reviews of Modern Physics* 51 (July 1979), 433–445.

176. S. Millman, ed., *A History of Engineering and Science in the Bell System: Physical Sciences (1925–1980)* (AT&T Bell Laboratories, 1983), pp. 269–274. A. A. Penzias and R. W. Wilson, "A Measurement of Excess Antenna Temperature at 4080 Mc/s," *Astrophysical Journal* 142 (1965), 419–421. Steven Weinberg, *The First Three Minutes: A Modern View of the Origin of the Universe* (New York: Basic Books, 1977).

177. On this general issue, see Everett Mendelsohn, Merritt Roe Smith, and Peter Weingart, "Science and the Military: Setting the Problem," introduction to *Science, Technology and the Military* (Amsterdam: Kluwer, 1988), xi–xxix. (I am grateful to M. R. Smith for showing me this article before publication.)

178. The Jet Propulsion Laboratory, run by California Institute of Technology and funded by the U.S. Army and NASA, also had a productive maser radar astronomy program.

179. Clayton Koppes makes a similar case—the largesse for space research in an era of missile warfare—in his *JPL and the American Space Program* (New Haven: Yale University Press, 1982). See also Leonard Reich's appreciation of early U.S. industrial laboratories in his book *The Making of American Industrial Research: Science and Business at GE and Bell, 1876–1926* (New York: Cambridge University Press, 1985): "One can hardly help but conclude that industrial research has, on balance, strongly promoted the pursuit of science and the advance of technology. . . [but] only so long as we understand that advance to be in the directions desired by major corporations" (pp. 8–9).

180. Christopher A. Hogg, Lawrence G. Suczy, et al., *Masers and Lasers* (Cambridge, Massachusetts: Maser/Laser Associates, 1962), pp. 177–178.

Chapter 3: The Birth of the Laser

1. Quarterly Report of the Columbia Radiation Laboratory, 31 March 1952, p. 11.

2. Edward I. Ginzton, "Microwaves—Present and Future," *IRE Transactions on Microwave Theory and Techniques* MTT-4 (1956), 136.

3. Paul D. Coleman, "State of the Art: Background and Recent Developments—Millimeter and Submillimeter Waves," *IEEE Transactions on Microwave Theory and Techniques* MTT-11 (September 1963), 271–288. See also Coleman and R. C. Becker, "Present State of the Millimeter Wave Generation and Technique Art—1958," ibid., MTT-7 (1959), 42–61.

4. I. Kaufman, "The Band Between Microwave and Infrared Regions," *Proc. IRE* 47 (1959), 381–396.

5. H. Motz, "Cerenkov and Undulator Radiation," *IRE Transactions on Antennas and Propagation* AP-4 (1956), 374–384.

6. Coleman, "State of the Art," cited in note 3, p. 278; H. H. Plotkin and F. H. Reder, "Meetings: Atomic Clocks and Microwave Amplification," *Physics Today* 9 (June 1956), 45.

7. Hughes Research Laboratory: H. Lyons, G. Birnbaum, and M. Stitch, "Atomic and Molecular Resonances for Millimeter-Wave Generation," *Third Quarterly Progress Report*, 1 January 1957–31 March 1957, SC Contract DA 36-039 SC-73080; also the *Final Report*, 1 July 1957 (courtesy of H. Lyons). *Proceedings of the First Tri-Service Millimeter Wave Symposium*, 10–11 September 1957, Army Signal Engineering Laboratories, Fort Monmouth, New Jersey; J. R. Singer, *IEEE Transactions on Microwave Theory and Techniques* MTT-7 (1959), 268–272. W. Gordy and M. Cowan, *Journal of Applied Physics* 31 (1960), 941–942; L. R. Momo et al., ibid., 443; S. Foner and L. R. Momo, ibid., 742–743; S. M. Bergman, ibid., 275–276. Chief, Microwave Tubes Branch, to Director of Engineering Operations, 8 May 1956, and I. R. Senitzky to Chief, Microwave Tubes Branch, 12 April 1957, Signal Corps Records, Federal Records Center, St. Louis, Missouri.

8. R. H. Dicke, "Notebook: Invention Disclosures, April 1945–May 1959." These ideas were elaborated into disclosures for RCA at the end of February, and a patent was filed by RCA attorneys in May 1956 and granted in 1958. See the file "RCA Disclosure 42,900. Infrared Oscillator and Amplifier Employing Circular Maser," R. H. Dicke Papers (courtesy of R. H. Dicke), and U.S. patent number 2,851,652, "Molecular Amplification and Generation Systems and Methods." A. M. Prokhorov published related suggestions in 1958: "Molecular Amplifier and Generator for Submillimeter Waves," *Zhurnal Eksp. i Teor. Fiz.* 34 (June 1958), 1658–1659, English translation in *Soviet Physics. JETP* 7 (December 1958), 1140–1141.

9. A. L. Schawlow, deposition, 24 June 1963, Interference #92,015 (courtesy of AT&T Legal Department); Irwin Wieder, private communications, 17 October 1985 and 12 November 1985; Rolf W. Landauer, interview, 17 October 1984 (SHL). See also R. S. Elliott, *Journal of Applied Physics* 23 (1952), 812.

10. W. J. Otting to J. L. Bromberg, 7 October 1983 (SHL); C. H. Townes, deposition, 1 July 1963, Interference #92,015 (courtesy of AT&T Legal Department); C. H. Townes to J. L. Bromberg, 26 February 1985, C. H. Townes Papers.

11. Motz, "Cerenkov and Undulator Radiation," cited in note 5. This work became the precursor of the free electron laser (discussed in the epilogue).

12. Robert H. Dicke, interview, 2 May 1983 (SHL).

13. This analysis draws heavily upon a critique of a previous draft of the book by C. H. Townes. At the same time, the interpretation given here remains quite different from his. See C. H. Townes to J. L. Bromberg, 31 July 1987, C. H. Townes Papers.

14. Senitzky to Chief, Microwave Tubes Branch, 12 April 1957, cited in note 7.

15. C. H. Townes and A. L. Schawlow, *Microwave Spectroscopy* (New York: McGraw-Hill 1955), p. 451.

16. R. W. Gelinas, "Masers and Irasers," Rand Corporation Report P-1585, 30 December 1958.

17. See also Mario Bertolotti, *Masers and Lasers: An Historical Approach* (Bristol: Adam Hilger, 1983), chap 5, and Bela A. Lengyel, "Evolution of Masers and Lasers," *American Journal of Physics* 34 (1966), 903–913.

18. C. H. Townes to Ron Shelton, 9 June 1986, C. H. Townes Papers.

19. The existence of this step in Townes's thinking is an inference made on the basis of his notebook, the letter quoted, and his article, "Quantum Optics or Quantum Electronics," in *Contemporary Physics: Trieste Symposium, 1968* (Vienna: International Atomic Energy Agency, 1969), vol. I, pp. 295–314. If the intensity of a spectral line is plotted as a function of its frequency, the line width is measured by the interval between the two frequencies at which the intensity falls to half its maximum value at line center.

20. Townes to Shelton, cited in note 18.

21. C. H. Townes, "Computation Book IV, July 1, 1957–May 1, 1965," pp. 15–23, dated 14 September 1957–28 October 1957, C. H. Townes Papers.

22. A. L. Schawlow, "Masers and Lasers," *IEEE Transactions on Electron Devices* ED-23 (July 1976), 776.

23. Arthur L. Schawlow, interview, 19 January 1984 (SHL).

24. C. H. Townes, "Computation Book IV," cited in note 21, pp. 17, 21.

25. Gordon Gould, "Notebook #1" (courtesy of R. G. Gould); R. G. Gould, deposition, 22 April 1963, Interference #92,015.

26. Gould, deposition, cited in note 25; Gordon Gould, interview, 10 April 1983, untranscribed tape, Niels Bohr Library, American Institute of Physics.

27. A. L. Schawlow to C. H. Townes, 3 April 1962, C. H. Townes Papers; A. L. Schawlow and C. H. Townes, "Infrared and Optical Masers," *Physical Review* 112 (1958), 1940–1949, and Schawlow and Townes, "Masers and Maser Communications System," U.S. Patent 2,929,922.

28. Schawlow and Townes, "Infrared and Optical Masers," cited in note 27, pp. 1941–1942. They noted, however, that there are a variety of effects that could push power requirements over this minimum.

29. Ibid, pp. 1947–1948.

30. L. A. Goldmuntz, interview, 21 October 1983 (SHL). See also Goldmuntz's testimony on behalf of Gould in Interference #92,015, 23 April 1963; "Company Profile: TRG Incorporated," *Microwave Journal* 4 (September 1961), 125–129; Robert Cushman, "Air Problems Attacked in Mid-Manhattan," *Aviation Week* 67 (8 July 1957), 99ff. (The company changed its name to TRG, Inc., when it incorporated in 1957.)

31. Cushman, "Air Problems," cited in note 30.

32. Richard Gordon Gould and Maurice Charles Neustein, depositions, 22 April 1963, and Alan Berman, deposition, 20 May 1963, Interference #92,015.

33. R. Gordon Gould, "Notebook #2," dated 28 August 1958 and after; TRG, Inc., "Proposal to Study the Properties of Laser Devices," 12 December 1958 (both courtesy of R. G. Gould). See Lengyel, "Evolution," cited in note 17. For a partisan, but useful, criticism of Gould's ideas, see Charles G. B. Garrett, deposition, 1 August 1963, Interference #92,015.

34. *New York Times*, 9 December 1965, p. 96.

35. R. J. Guenther, A. J. Torsiglieri, and M. Braunstein, "Brief on Behalf of Schawlow and Townes," 16 December 1963, Interference #92,015, p. 70.

36. H. G. J. Aitken, *The Continuous Wave: Technology and American Radio, 1900–1932* (Princeton: Princeton University Press, 1985), p. 548.

37. T. S. Kuhn, in Marshall Clagett, ed., *Critical Problems in the History of Science* (Madison, Wisconsin: University of Wisconsin Press, 1959), p. 323.

38. Thomas S. Kuhn, "Historical Structure of Scientific Discovery," *Science* 136 (1 June 1962), 760–764, reprinted in *The Essential Tension* (Chicago: University of Chicago Press, 1977), pp. 165–177. See also T. S. Kuhn, *The Structure of Scientific Revolutions* (Chicago: University of Chicago Press, 1962).

39. See Lengyel, "Evolution," cited in note 17, pp. 909–910.

40. Townes, deposition, cited in note 10.

41. C. H. Townes, transcript of an interview by C. Breck Hitz, 27 September 1984, for *Lasers and Applications* (courtesy of C. H. Townes). A slightly altered version appears in *Lasers and Applications: Laser Pioneer Interviews* (Torrance, California: High Tech Publications, 1985), p. 41.

42. C. H. Townes, "Proposal for Research on a Maser to Amplify or Oscillate at Infrared Frequencies, September 1, 1958 through August 31, 1959," 9 July 1958, and "December 1, 1959 through November 30, 1960," 15 June 1959, Papers of the Columbia University Physics Department, Columbia University Archives.

43. Otting to Bromberg, cited in note 10; folder, "Contract AF 49[683]-507. Correspondence," Papers of the Columbia University Physics Department, Columbia University Archives.

44. Schawlow, "Masers and Lasers," cited in note 22, pp. 777–778; and "From Maser to Laser," in B. Kursunoglu and A. Perlmutter, eds., *Impact of Basic Research on Technology* (New York: Plenum Press, 1973), pp. 113–148.

45. C. H. Townes, "Memo to members of the Steering Committee of the coming conference on quantum electronics, December 5, 1958," and C. H. Townes to Samuel Sensiper, 23 February 1959, C. H. Townes Papers.

46. Ali Javan, deposition, 1 July 1963, and A. L. Schawlow, deposition, 24 June 1963, Interference #92,015 (courtesy of AT&T Legal Department); Ali Javan, deposition, 16 February 1968, Interference #94,835 (courtesy of Patlex Corp.).

47. Ali Javan, interview by Jeff Hecht, *Lasers & Applications* (October 1985), 49–53. On the disadvantages of optical pumping, see William R. Bennett jr., "Gaseous Optical Masers," *Applied Optics*, supplement. 1 (1962), 35.

48. Javan, deposition, 1968, cited in note 46.

49. J. H. Sanders to R. Kompfner, 10 January 1958 (Box 6) and 4 November 1958 (Box 7), and R. Kompfner to J. H. Sanders, 17 October 1958 (Box 7), R. Kompfner Papers, AT&T Archives; J. H. Sanders, "Optical Maser Design," *Physical Review Letters* 3 (1959), 86–87; Gary D. Boyd, paraphrase of an interview for the Laser History Project, 1 May 1986 (SHL).

50. Javan, deposition, 1968, cited in note 46.

51. Goldmuntz, deposition, cited in note 30.

52. This, of course, is a forerunner of the Strategic Defense Initiative, or Star Wars program, that Ronald Reagan initiated during the 1980s. See the epilogue and also the "Report to the American Physical Society of the Study Group on Science and

Technology of Directed Energy Weapons," *Reviews of Modern Physics* 59, no. 3, part 2 (July 1987, supplement), S1–S201.

53. "Information Sheet for Prospective Bidders on ARPA's 'Guide Line Identification Program for Anti-missile Research' (GLIPAR), Phase I," 28 January 1959, R. Kompfner Papers, Box 6, AT&T Archives; George F. Smith, interview, 5 February 1985 (SHL); TRG, Inc., "Proposal to Study the Properties of Laser Devices," 12 December 1958; R. W. Seidel, "How the Military Responded to the Laser," *Physics Today* 41 (October 1988), 12–13.

54. Goldmuntz, deposition, cited in note 30; contract AF49[638]-673 between AFOSR and TRG, Inc. (courtesy of R. G. Gould).

55. R. Gordon Gould, interview, 23 October 1983 (SHL); Goldmuntz, interview, cited in note 30; Richard T. Daly, interview by Paul Forman, 28 May 1984, untranscribed tape, Niels Bohr Library, American Institute of Physics; Stephen Jacobs, memoir for the Laser History Project, 1984 (SHL); Paul Rabinowitz, deposition, 23 May 1968, Interference #94,837 (courtesy of Patlex Corp.).

56. *The Ann Arbor Conference on Optical Pumping*, University of Michigan, 15–18 June 1959, pp. 128–138 (courtesy of the Research Library, Westinghouse Research and Development Center).

57. A "narrow" line is one with a very small frequency spectrum. In this discussion, I have conflated the ideas of spectral line and energy level, a practice endemic in the laser literature. If a molecular system has a narrow energy level (one that stretches over a small band of energies), then, when it makes a transition to another narrow level, it emits a narrow spectral line.

58. *Ann Arbor Conference*, cited in note 56, pp. 133–136; Quarterly Technical Report, No. 4 (February–June 1958), No. 5 (June–September 1958), No. 6 (September–November 1958), and No. 8 (February 15–May 15 1959), Wright Air Development Center Contract AF33(616)-5258-R8, Westinghouse Research Laboratories (courtesy of I. Wieder).

59. Interviews with Joseph A. Giordmaine, 31 May and 4 June 1984 (SHL). B. Lax, "Conference on Quantum Electronics," *Microwave Journal* 3 (February 1960), 21ff.

60. Isaac D. Abella, paraphrase of an interview for the Laser History Project, 22 October 1985 (SHL). The work can be followed in the notebooks of Cummins and Abella and in the correspondence of O. S. Heavens, all in the Columbia University Physics Department Papers, Columbia University Archives.

61. Schawlow, "From Maser to Laser," cited in note 44, and "Infrared and Optical Masers," in C. H. Townes, ed., *Quantum Electronics: A Symposium* (New York: Columbia University Press, 1960), pp. 553–563.

62. Robert Sobel, *IBM, Colossus in Transition* (New York: Times Books, 1981), pp. 159–166; M. B. W. Graham, "Industrial Research in the Age of Big Science," in Richard Rosenbloom, ed., *Research on Technological Innovation, Management and Policy*, vol. 2 (Greenwich, Connecticut: Jai Press, 1985), pp. 47–79; *New York Times*, 22 September 1956, p. 25, 27 September 1956, p. 58, 31 December 1956, p. 10.

63. W. V. Smith to J. L. Bromberg, October 1983 and 25 September 1985 (SHL).

64. P. P. Sorokin, "Contributions of IBM to Laser Science—1960 to the Present," *IBM Journal of Research and Development* 23 (1979), 476–488.

65. P. P. Sorokin, interview by Jeff Hecht, 27 September 1984 (SHL).

66. "Temperature and Concentration Effects in a Ruby Maser," in Townes, ed., *Quantum Electronics,* cited in note 61, pp. 324–332.

67. Theodore H. Maiman, deposition, 30 November 1967, Interference #94,837, and Hughes Aircraft Company Research Laboratories, "Annual Status Report, 1 July 1958 through 30 June 1959" (both courtesy of G. F. Smith).

68. Irnee J. D'Haenens, interview, 5 February 1985 (SHL).

69. T. H. Maiman, "Generation of Infrared and Optical Radiation by Maser Techniques," manuscript, 11 December 1959 (courtesy of I. J. D'Haenens); Maiman, deposition, cited in note 67.

70. Maiman, deposition, cited in note 67.

71. Javan, deposition, 1968, cited in note 46; William R. Bennett jr., deposition, 29 February 1968, Interference #94,835 (courtesy of Patlex Corp.); Donald R. Herriott to J. L. Bromberg, 23 November 1983 and 23 October 1985 (SHL); W. R. Bennett, interview, 26 October 1987 (SHL).

72. See Anthony E. Siegman, *Lasers* (Mill Valley, California: University Science Books, 1986), pp. 43–49.

73. See, for example, W. E. Lamb jr., "Laser Theory and Doppler Effects," *IEEE Journal of Quantum Electronics* QE-20 (1984), 552.

74. A. G. Fox and Tingye Li, "Resonant Modes in a Maser Interferometer," *Bell System Technical Journal* 40 (1961), 453–488.

75. Snitzer was limited to unclassified research at this time. He had refused to testify before the House UnAmerican Committee in March 1958 on the grounds of his First Amendment right to freedom of speech and association, and he had been fired by Lowell Technological Institute in an action that the American Association of University Professors ruled "grossly violated those principles of this Association which embody the concept of due process." See "Academic Freedom and Tenure: Lowell Technological Institute," *AAUP Bulletin* 45 (1959), 550–567, quote on p. 567. Also see Elias Snitzer, interview, 6 August 1984 (SHL).

76. Snitzer, interview, cited in note 75.

77. I. J. D'Haenens, "History of Maser Research and Development at Hughes Aircraft Company," typescript, 4 pp., 11 August 1983 (SHL); Maiman, deposition, cited in note 67; George F. Smith, "Early Laser Years at Hughes Aircraft Company," *IEEE Journal of Quantum Electronics* QE-20 (1984), 578.

78. Peter A. Franken, interview, 8 March 1985 (SHL).

79. Maiman, deposition, cited in note 67; George L. Trigg to J. L. Bromberg, 28 August 1984 (SHL). It is not clear whether Maiman's article was sent to an outside referee or refereed informally by the journal staff (see Trigg to Bromberg). The grounds for rejection are also unclear. Goudsmit had a policy against publishing maser papers that did not "contain significant contributions to basic physics" (see his editorial in *Physical Review Letters* 3 [1959], p. 125) and also against publishing serial articles from the same investigation (Maiman had just published his experiments on the quantum efficiency of ruby in the 1 June 1960 issue).

80. "Stimulated Optical Radiation in Ruby," *Nature* 187 (6 August 1960), 493–494.

81. Maiman describes a meeting in July 1960 at the Institute for Defense Analyses as an example of this skepticism; see his deposition, cited in note 67.

82. R. J. Collins, D. F. Nelson, A. L. Schawlow, W. Bond, C. G. B. Garrett, and W. Kaiser, "Coherence, Narrowing, Directionality, and Relaxation Oscillations in the Light Emission from Ruby," *Physical Review Letters* 5 (1 October 1960), 303–305. Schawlow discusses the history of this research in "From Maser to Laser," cited in note 44.

83. An example of this is a letter by D. Lainé, *Nature* 191 (19 August 1961), 795–796; See also *Time* 76 (17 October 1960), 47–48.

84. Snitzer, interview, cited in note 75.

85. T. H. Maiman, "Optical and Microwave-Optical Experiments in Ruby," *Physical Review Letters* 4 (1960), 564–566.

86. Irwin Wieder, private communication, 12 November 1985; I. Wieder and L. Sarles, "Stimulated Optical Emission from Exchange-Coupled Ions of Cr^{+++} in Al_2O_3," *Physical Review Letters* 6 (1961), 95–96; A. L. Schawlow and G. E. Devlin, "Simultaneous Optical Maser Action in Two Ruby Satellite Lines," ibid., 96–98.

87. Herriott to Bromberg, 23 November 1983, cited in note 71.

88. Javan, deposition, 1968, cited in note 46; Herriott to Bromberg, cited in note 71; Bennett, interview, cited in note 71; A. Javan, W. R. Bennett jr., and D. R. Herriott, "Population Inversion and Continuous Optical Maser Oscillation in a Gas Discharge Containing a He-Ne Mixture," *Physical Review Letters* 6 (1961), 106–108.

89. R. W. Seidel has pointed out that Bell Laboratories was also heavily involved in radar, which requires pulsed generators. I believe, though, that it was AT&T's communications function that loomed largest for the Research Area staff.

90. Sidney Millman, interview, 5 June 1984 (SHL).

91. Townes indicated in a private telephone conversation in October 1985 that he does not think this interpretation catches the essentials of the situation. Each investigator aimed at making a working system, in the full confidence that other lasers, with other properties, would follow. In his opinion, differences in the initial system they chose to explore were consequences of style and background, but not of institution. Townes further recalled that he and other laser pioneers fully expected a wide variety of applications for the laser.

Chapter 4: Laser Research Takes Off

1. N. Bloembergen, "Introduction," in P. Grivet and N. Bloembergen, eds., *Quantum Electronics: Proceedings of the Third International Congress, Paris* (New York and Paris: Columbia University Press and Dunod, 1964); J. R. Singer, *Advances in Quantum Electronics* (New York: Columbia University Press, 1961), 631–641.

2. Edward V. Ashburn, ed., *Laser Literature: A Permuted Bibliography, 1958–1966* (North Hollywood, California: Western Periodicals, 1967), p. vi.

3. Alan D. White, paraphrase of an interview for the Laser History Project, 30 April 1986 (SHL).

4. Marshall I. Nathan, interview, 17 October 1984 (SHL).

5. Paul Forman, "Behind Quantum Mechanics," *Historical Studies in the Physical and Biological Sciences* 18(1) (1987), 170. Forman estimates that for the 1950s, the total number of unclassified reports was roughly equal to the total of published papers, and the number of company-proprietary or security-classified reports was probably significantly larger. Data from *Physics Abstracts* probably underrepresent government laboratories for the same reasons.

6. Barry Miller, "Services to Push Optical Maser Effort," *Aviation Week & Space Technology* 76 (15 January 1962), 92ff. A Harvard Business School case study estimated that the number of active research groups went from roughly 25 or 50 in both the United States and Europe in 1960 to more than 500 in the United States alone less than two years later: Christopher A. Hogg, Lawrence G. Sucsy, et al., *Masers and Lasers* (Cambridge, Massachusetts: Maser/Laser Associates, 1962), pp. 178–179.

7. See the section on the laser industry below. Nonlinear optics makes use of the fact that laser light is so intense that not only the usual effects that are linear in the electric field strength, but also those of higher order, are significant enough to be observed.

8. L. D. Smullin and G. Fiocco, "Project Luna See," *Proc. IRE* 50 (1962), 1703–1704.

9. Rikard Stankiewicz, *Academics and Entrepreneurs: Developing University-Industry Relations* (New York: St. Martin's Press, 1986), pp. 44–49, estimates that over 60% of U.S. engineering faculty and about 30% of physical science faculty were involved in consultancies in the 1980s. However, in conversations with 15 academic laser physicists, engineers, and chemists active in the 1960s, I have found that 12 held consultancies of some type in those years. The remaining three were not asked about their consultancies.

10. See the interviews with William R. Bennett jr., 26 October 1987, and Roy J. Glauber, 21 April 1987 (SHL).

11. Robert W. Seidel, "From Glow to Flow: A History of Military Laser Research and Development," *Historical Studies in the Physical and Biological Sciences* 18(1) (1987), 111–147 and "How the Military Responded to the Laser," *Physics Today* 41 (October 1988), 12–19.

12. Elias Snitzer, interview, 6 August 1984 (SHL); E. Snitzer, "Optical Maser Action of Nd^{+3} in a Barium Crown Glass," *Physical Review Letters* 7 (1961), 444–446. Earlier, Leo F. Johnson at Bell Telephone Laboratories lased neodymium in the crystal host calcium tungstate ($CaWO_4$), a result that gave Snitzer further encouragement in pursuing the direction he had chosen.

13. Seidel, the two works cited in note 11.

14. *Aviation Week* 73 (5 December 1960), 97, gives almost $1.4 million for Air Force funding alone. The figure of $4 million is calculated by subtracting $1 million from the "estimated $5 million spent by military agencies to date" in Miller, "Services to Push Optical Maser Effort," cited in note 6.

15. J. Richard Elliott jr., "Spotlight on Lasers: A Brilliant Scientific Achievement Is Far From a Commercial Payoff," *Barron's* (25 February 1963), 3ff., and *Aviation Week and Space Technology* 78 (22 April 1963), 54.

16. If we eliminate from consideration the six papers from Lincoln Laboratory, and the two from Townes at the Institute for Defense Analyses (listed as university

papers because of Townes's Columbia affiliation), the numbers are 33 out of 44. Six of the remaining 11 had AEC or NASA support.

17. Barry Miller, *Aviation Week and Space Technology* 76 (12 March 1962), 229; 78 (22 April 1963), 54ff.

18. Harland Manchester, "Light of Hope—or Terror?" *The Reader's Digest* 82 (February 1963), 98.

19. Elliott, "Spotlight," cited in note 15; *Business Week* (18 August 1962), 62.

20. The APS figure is taken from *Physics Today* 21 (July 1968), 79, the OSA figure (it was 2,641 at the beginning of 1960 and 2,901 in December) is from the OSA membership office and from the minutes of the 1961 meetings of the Board of Directors (Director's files, OSA), and the IRE figure, valid as of 31 May 1959, is from the *1960 IRE Directory*, p. 3.

21. 50th Year History Committee (Hilda G. Kingslake, chair), "History of the Optical Society of America, 1916–1966," *JOSA* 56 (1966), 273–340; *JOSA*, passim, January 1960–June 1961.

22. Van Zandt Williams, "Optics—An Action Program," *JOSA* 53 (June 1963), 772–774. The 7.8% figure is for 1960, the 1.5% one for June 1962. See also National Research Council Physics Survey Committee, *Physics in Perspective* (Washington, D.C.: National Academy of Sciences, 1972), vol. I, pp. 184–200.

23. Mario Bertolotti, *Masers and Lasers: An Historical Approach* (Bristol: Adam Hilger, 1983), pp. 203–217; Robert Hanbury Brown, *The Intensity Interferometer* (London: Taylor and Francis, 1974), chapters 1 and 2; Max Born and Emil Wolf, *Principles of Optics: Electromagnetic Theory of Propagation, Interference and Diffraction of Light* (New York: Pergamon, 1959; sixth edition, 1980), chapters 7 and 10 in 3rd and subsequent editions.

24. William S. Rodney, "Item of the Week: Photonics," 20 April 1959, C. H. Townes Papers.

25. R. and H. G. Kingslake, "A History of the Institute of Optics," *Applied Optics* 9 (1970), 789–796.

26. Born and Wolf, *Principles of Optics*, cited in note 23.

27. Interviews with Emil Wolf and Robert E. Hopkins, 23 September 1984 (SHL); "Report to Advisory Panel [of the Institute of Optics]" ca. 1962, folder, "National Institute of Modern Optics, 1961–1962," E. Wolf Papers; minutes of the meeting of the Board of Directors, OSA, 3 February 1960 (OSA Director's files).

28. See the first two articles cited in note 66, below.

29. "Coherence Properties of Electromagnetic Radiation; Report on Conference Held at the University of Rochester . . .," AFOSR-583. Technical Note No. 5, Contract No AF 49 (638)-602, April 1961 (courtesy of E. Wolf); E. L. O'Neill and L. C. Bradley, "Coherence Properties of Electromagnetic Radiation," *Physics Today* 14 (1961), 28–34.

30. L. Mandel, "Photon Degeneracy in Light from Optical Maser and Other Sources," *JOSA* 51 (1961), 797–798.

31. Wolf, interview, cited in note 27. Wolf also had financial support from the Army Research Office.

32. Monochromaticity is the frequency purity of a single signal, whereas coherence "necessarily refers not to one property of a signal at a single point in space and time, but to a *relationship*, or *a family of relationships*, between one signal at one point in space and time, and the same or another signal at other points in space and time." See Anthony E. Siegman, *Lasers* (Mill Valley, California: University Science Books, 1986), p. 54.

33. L. Mandel and E. Wolf, "Some Properties of Coherent Light," *JOSA* 51 (1961), 815–819, and "The Measures of Bandwidth and Coherence Time in Optics," *Proceedings of the Physical Society* 80 (1962), 894–897.

34. E. Wolf, "Spatial Coherence of Resonant Modes in a Maser Interferometer," *Physics Letters* 3 (1963), 166–168; for a more recent treatment, see Emil Wolf and G. S. Agarwal, "Coherence Theory of Laser Resonator Modes," *JOSA* A (May 1984), pp. 541–546. See also O. S. Heavens, "Optical Masers," *Applied Optics,* supplement 1 (1962), 1–23, esp. 2–4. For a review of research in the field of coherence up to about 1965, see L. Mandel and E. Wolf, "Coherence Properties of Optical Fields," *Reviews of Modern Physics* 37 (1965), 231–287; for some of the original papers, see L. Mandel and E. Wolf, eds., *Selected Papers on Coherence and Fluctuations of Light,* vol. I: 1850–1960, and vol. II: 1961–1966 (New York: Dover, 1970).

35. Glauber, interview, cited in note 10; E. M. Purcell, "The Question of Correlation between Photons in Coherent Light Rays," *Nature* 178 (1956), 1449–1450; G. A. Rebka and R. V. Pound, "Time-Correlated Photons," *Nature* 180 (1957), 1035–1036.

36. One summary of Glauber's 1960s coherence work is given in his "Quantum Theory of Coherence," in S. M. Kay and Arthur Maitland, eds., *Quantum Optics* (New York: Academic, 1970), pp. 53–125. These 'functions' are expressions involving quantum-mechanical operators.

37. Bloembergen and Grivet, eds., *Quantum Electronics,* cited in note 1, vol. I, pp. 3–200.

38. See Roy J. Glauber, "Photon Correlations," *Physical Review Letters* 10 (1963), 84; Glauber, interview, cited in note 10; Wolf, interview, cited in note 27; L. Mandel and E. Wolf, "Photon Correlations," *Physical Review Letters* 10 (1963), 276–277; E. C. G. Sudarshan, "Equivalence of Semiclassical and Quantum Mechanical Descriptions of Statistical Light Beams," *Physical Review Letters* 10 (1963), 277–279.

39. "Program of the 1961 Spring Meeting of the Optical Society of America," *JOSA* 51 (1961), 471ff.; "Editor's Page," ibid., 490.

40. P. A. Franken, interview, 8 March 1985 (SHL).

41. Minutes of the meetings of the Board of Directors, OSA, for 13 March 1962, 2 October 1962, 24 March 1963, and 22 October 1963 (Director's files, OSA); John Howard, private communication, 30 July 1982.

42. A. L. Schawlow, interview for the Laser History Project and Stanford University, 19 January 1984 (SHL); OSA Committee Lists, *JOSA* 52 (February 1962), 226–230.

43. Hilda G. Kingslake, *The First Fifty Years: The Institute of Optics, 1929–1979* (Rochester: University of Rochester, College of Engineering and Applied Science, n.d.).

44. A. Michal McMahon, *The Making of a Profession: A Century of Electrical Engineering in America* (New York: IEEE Press, 1984), pp. 216–217, 239–243.

45. See Joan Lisa Bromberg, "Engineering Knowledge in the Laser Field," *Technology & Culture* 27 (1986), 813–816. After the merger, the PGs became PTGs, Professional Technical Groups.

46. Minutes of the Administrative Committee of the IRE/IEEE Professional Group on Electron Devices, 20 May 1963 (IEEE Headquarters, New York).

47. Minutes of the Administrative Committee of the IRE/IEEE Professional Group on Microwave Theory and Techniques, 25 March, 20 May, and 11 September 1963 (IEEE Headquarters, New York).

48. PGED and PGMTT Administrative Committee Minutes, 1960–1964, passim (IEEE Headquarters, New York).

49. PTGMTT Minutes, 30 November 1964, and PTGED Minutes, 26 August 1964 (IEEE Headquarters, New York).

50. Eugene I. Gordon, interview, 3 June 1984 (SHL).

51. Minutes, passim, 1965 (IEEE Headquarters, New York). The Quantum Electronics Council metamorphosed in 1977 into the IEEE Quantum Electronics and Applications Society, the independent group that some IRE/IEEE members had sought in the early 1960s.

52. Gordon, interview, cited in note 50.

53. Eighteen firms are listed for 1963 in the "Laser Family Tree" generated from the recollections of a panel of 15 industry veterans and printed in *Lasers & Applications* (July 1985). There are 29 firms listed under the headings for light amplifiers, and light oscillators, by stimulated emission ("lasers" and "losers") that also list lasers or losers among their products in the suppliers section of the *Optical Industry and Systems Directory 9* (Lenox, Massachusetts: Optical Pub. Co., July 1963).

54. Elliott, "Spotlight," cited in note 15, pp. 3ff. The market analysis firm of Technology Markets, Inc., estimated in mid-1962 that lasers, masers, and related equipment all together would not amount to $10 million in sales in that year; see *Business Week* (18 August 1962), 62.

55. EIA, *Electronics Industries Year Book*, 1965, p. 2.

56. John E. Tilton, *International Diffusion of Technology: The Case of Semiconductors* (Washington, D.C.: Brookings Institution, 1971); *Fortune* 68 (November 1963), 137.

57. *Electronics* 34 (27 January 1961), 14, estimated that 12% of each shipbuilding dollar, 25% of each aircraft dollar, 35% of each missile dollar, and 30% of each federal R&D dollar went into electronics; see also Herman O. Stekler, *The Structure and Performance of the Aerospace Industry* (Berkeley: University of California Press, 1965), p. 22. For the Kennedy military buildup, see S. E. Ambrose, *Rise to Globalism: American Foreign Policy Since 1938* (London: Penguin Books, 1971).

58. *Electronics* 34 (9 June 1961), 14. For the electronics industry's optimistic assessment of its prospects, look through the issues of *Electronics* for January–June 1961. For Kennedy's change of plans, see the *New York Times Index* for 1961, "Astronautics," and John Logsden, *The Decision to Go to the Moon: Project Apollo and the National Interest* (Cambridge: The MIT Press, 1970).

59. Robert C. Rempel, interview, 18 January 1984, untranscribed tape, Niels Bohr Library, American Institute of Physics.

60. Battelle Memorial Institute, Columbus, Ohio, *The Implications of Reduced Defense Demand for the Electronics Industry* (U.S. Arms Control and Disarmament Agency, September 1965), quotation on p. 39; Frank R. Bacon jr. and Katherine A. Rempp, *Electronics in Michigan: A Study of the Electronics Industry in Michigan in Its National and International Setting* (Ann Arbor: Institute of Science and Technology, University of Michigan, 1967); Arthur M. Johnson, "The Economy Since 1914" in Glenn Porter, ed., *Encyclopedia of American Economic History* (New York: Scribner, 1980), vol. 1, pp. 110–130; *Electronics*, passim, 1961; Thomas C. Cochran, *200 Years of American Business* (New York: Basic, 1977).

61. Trion was one of the smaller of the high-technology firms that began to spring up around the University of Michigan starting in the mid-1950s. See Michael Wolff, "The Birth of Holography, I: A New Process Creates an Industry," *Innovation Magazine* no. 7 (1969), 4–15. Lee A. Cross, 10 May 1984, interview, and Donald E. Gillespie and Chihiro Kikuchi, paraphrase of an interview, 19 October 1985 (SHL).

62. "Trion Instruments, Inc. Schedule of Product Sales, March 1, 1961 thru November 30, 1961," folder "Lear Siegler [Trion Instruments, 1961–1963]," P. A. Franken Papers. For Trion's status as a leading firm, see Hogg, Sucsy, et al., *Masers and Lasers*, cited in note 6.

63. Interviews with Elsa M. Garmire, 4 February 1985, and Michael Bass, 29 May 1985 (SHL).

64. Franken, interview, cited in note 40.

65. The vibration, or oscillating electric polarization P is related to the incident field E through $P = E + aE^2 + bE^3 + \ldots$, where the coefficients are only about 10^{-9}, so that terms higher than E only have an effect for very large E. The calculations are in a manuscript, dated 11 April 1961, in the folder "Laser Harmonics," P. A. Franken papers; see also Peter Franken, "High-Energy Lasers," *International Science and Technology* (October 1962), 62–68, esp. 67.

66. For Forrester's work see A. T. Forrester, W. E. Parkins, and E. Gerjuoy, "On the Possibility of Observing Beat Frequencies between Lines in the Visible Spectrum," *Physical Review* 72 (1947), 728, and A. T. Forrester, R. A. Gudmundsen, and P. O. Johnson, "Photoelectric Mixing of Incoherent Light," *Physical Review* 99 (1955), 1691–1700. For examples of laser proposals, see Forrester, "Photoelectric Mixing as a Spectroscopic Tool," *JOSA* 51 (1961), 253–259, and D. C. Lainé, "A Proposal for a Tunable Source of Radiation for the Far Infra-red using Beats between Optical Masers," *Nature* 191 (19 August 1961), 795–796.

67. Franken, interview, cited in note 40.

68. The "head" is the laser tube and excitation source (flash lamps in the case of the ruby laser), decoupled from the power supply.

69. In the printed article, the illustration, which shows an arrow pointing at nothing, is captioned, in part, "the arrow at 3472 Å indicates the small but dense image produced by the second harmonic"; see P. A. Franken, A. E. Hill, C. W. E. Peters, and G. Weinreich, "Generation of Optical Harmonics," *Physical Review Letters* 7 (1961), 118–119.

70. Peter A. Franken to Douglas L. Linn, 20 June 1961, in the folder "Lear Siegler [Trion Instruments] 1961–1963," P. A. Franken Papers.

71. G. C. Dacey, "Optical Masers in Science and Technology," *Science* 135 (12 January 1962), 74.

72. Franken, "High-Energy Lasers," cited in note 65, pp. 62–23; Lee Cross, interview, cited in note 61. Robert C. Rosan, Mary K. Healy, and William F. McNary jr., "Spectroscopic Ultramicroanalysis with a Laser," *Science* 142 (11 October 1963), 236–237. There was some controversy over the relative contributions of Franken and Brech; I use the version in "Statement by Peter Franken. . . with which Fred Brech concurs, October 11, 1962," in the folder "Lear Siegler [Trion instruments, 1961–1963]," P. A. Franken Papers.

73. The discussion of Spectra-Physics is based upon the following sources: Interviews with Herbert M. Dwight jr., Robert C. Rempel, and Kenneth A. Ruddock for the Laser History Project, 18 January 1984 (a transcript of the Dwight interview is in SHL; untranscribed tapes of the other interviews are at the Niels Bohr Library, American Institute of Physics); Gene Bylinsky, "Bringing the Laser Down To Earth," *Fortune* 80 (September 1969), 126–129; Scrapbooks of news clippings in the records of Spectra-Physics (courtesy of H. M. Dwight); British Patent 984,590; U.S. Patents 3,333,101, and 3,204,471; Arnold L. Bloom, interview by Paul Forman, 1 August 1983 (SHL).

74. John F. Kreidl, "The Optics Industry Acquires a New Look," part 2, *Laser Focus* 3 (January 1967), 20–22.

75. *Palo Alto Times*, 14 May 1963 (courtesy of W. E. Bell).

76. G. C. Dacey to A. W. Horton jr., "Visit to Assess Patent Potential of the Perkin-Elmer Corporation," 17 May 1961, Case 38788-1, AT&T Archives.

77. The discovery of the red helium-neon line is discussed in the section on gas masers in this chapter.

78. Charles Freed, private communication, 3 March 1988.

79. Fred P. Burns, paraphrase of an interview, 30 March 1987 (SHL). This interview is the source of most of the information on Korad in this section.

80. For some interesting asides on the military's relations with laser companies, see Malcolm L. Stitch, "Evolution of a Trade Association," *Space Age News* (October 1968), 9.

81. See Battelle, *Implications of Reduced Defense Demand*, cited in note 60, p. 37.

82. Bernard H. Soffer, paraphrase of an interview for the Laser History Project, 29 May 1985 (SHL).

83. A. L. Bloom, W. E. Bell, and R. C. Rempel, "Laser Operation at 3.39 Microns in a Helium-Neon Mixture," *Applied Optics* 2 (1963), 317–318. Q-switching is discussed in the next section. For ion lasers, see chapter 5.

84. Bylinsky, "Bringing the Laser Down to Earth," cited in note 73.

85. Burns, paraphrase of an interview, cited in note 79.

86. Reported in *Business Week* (18 August 1962), 62. See also the paraphrases of interviews with Colin Bowness, 9 August 1985, and A. Stevens Halsted, 1 December 1985 (SHL).

87. Harry E. Franks to J. L. Bromberg, 8 May 1987 (SHL).

88. "Company Profile: Optics Technology...," *Laser Focus Report* (15 August 1966), 7.

89. "Maser Patent Correspondence: Research Corporation" for 1961–1963 and 1964–1966, C. H. Townes papers; also, "What's a Patent Worth?" *Electronics* 38 (20 September 1965), 137ff.

90. Schawlow and Townes won in the Patent Office in November 1964. Patent litigation, however, still continues.

91. "What's a Patent Worth?," cited in note 89.

92. The decree is reprinted in *Hearings before the Antitrust Subcommittee of the Committee on the Judiciary, House of Representatives, 85th Congress, second session, 1958*, part II, vol. II. See also John Brooks, *Telephone: The First Hundred Years* (New York: Harper & Row, 1976).

93. "What's a Patent Worth?," cited in note 89.

94. Information supplied by Hughes Aircraft Company Archives.

95. Among other U.S. groups working on radar early on were RCA's Aerospace Division, TRG, Martin-Marietta's Orlando Division, Westinghouse, Sperry, The Frankford Arsenal, and the Army Signal Corps Electronics Research and Development Laboratories at Fort Monmouth, New Jersey.

96. *Electronics* 40 (27 November 1967), 48.

97. G. F. Smith to A. V. Haeff, 21 July 1960 (courtesy of G. F. Smith).

98. Information supplied by Hughes Aircraft Company Archives.

99. George F. Smith, private communication, 3 November 1987; Douglas A. Buddenhagen, private communication, 9 December 1987; information from Hughes Aircraft Company Archives.

100. *Hughesnews* (11 May 1962), Hughes Aircraft Company Archives.

101. For discussions of entrepreneurial groups within large corporations, see Edward B. Roberts, "A Basic Study of Innovators," *Research Management* 11 (1968), 249–266; Mel Horwitch and C. K. Prahalad, "Managing Technological Innovation—Three Ideal Modes," *Sloan Management Review* 17 (winter 1976), 77–89.

102. *Hughesnews*, cited in note 100; *Business Week* (12 May 1962),78; information from Hughes Aircraft Company Archives; Buddenhagen, private communication, cited in note 99.

103. See "Raytheon Company," in *Moody's Industrial Manual* for 1961. Raytheon shifted its product mix toward the civilian side as the decade advanced.

104. M. Ciftan et al., "A Ruby Laser with an Elliptic Configuration," *Proc. IRE* 49 (1961), 960–961. There were disadvantages to the linear lamp vs. the spiral lamp configuration, however, and Maiman's company, Korad, stayed with the spiral lamp configuration. See Burns, paraphrase of an interview, cited in note 79.

105. Colin Bowness, paraphrase of an interview for the Laser History Project, 9 August 1985 (SHL); *Pulse* (an in-house publication of Raytheon's Microwave and Power Tube Division) 3(2) (March 1961); Special Microwave Devices Operation, "Catalog of Microwave Devices, Materials, and Lasers," n.d. (ca. 1962); T. L.

Phillips to Division General Managers, "Raytheon Laser Program," 24 August 1962 (all courtesy of C. Bowness).

106. H. Statz, G. A. deMars, A. Adams jr., D. T. Wilson, and J. W. Barnie, private communication, 16 March 1988.

107. David Bushnell, "Laser Applications: Ground Systems," *Electronics Progress* (Raytheon in-house publication) 7(4) (1963), 21–23.

108. C. L. Tang, H. Statz, and G. A. deMars, "Regular Spiking and Single-Mode Operation of Ruby Laser," *Applied Physics Letters* 2 (1963), 222–224. Another case in point is the article "Zeeman Effect in Gaseous Helium-Neon Optical Masers" by H. Statz, Roy A. Paananen, and G. F. Koster, *Journal of Applied Physics* 33 (1962), 2319–2321, which contained a calculation of how the energies of the upper and lower laser levels change in the presence of a magnetic field. This could be technologically interesting because it meant that a magnetic field might be used to frequency-modulate the laser output for communications systems: See *Electronics Progress* 7(4) (1963), 23.

109. T. L. Phillips to Division General Managers, 24 August 1962, including "Proposed Charter for Lasers," cited in note 105.

110. J. H. Dunnington, E. F. Regan, and F. A. L'Esperance jr., "Progress in Ophthalmic Surgery," part 2, *New England Journal of Medicine* 269 (1963), 407–413.

111. *Wall Street Journal* (22 December 1961), 5

112. Charles J. Koester, memoir for the Laser History Project, n.d. (ca. 1985) (SHL); Charles J. Campbell et al., "Intraocular Temperature Changes Produced by Laser Coagulation," *Acta Ophthalmologica Supplement* 76 (1963), 22–31. Other firms that placed photocoagulators on the market in these years, such as Optics Technology and TRG, also developed their instruments through a collaboration between scientists and physicians. For some of the early collaborative research, see N. S. Kapany, N. A. Peppers, H. C. Zweng, and M. Flocks, "Retinal Photocoagulation by Lasers," *Nature* 199 (1963), 146–149; Leonard R. Solon, Raphael Aronson, and Gordon Gould, "Physiological Implications of Laser Beams," *Science* 134 (10 November 1961), 1506–1508; Milton M. Zaret et al., "Ocular Lesions Produced by an Optical Maser (Laser)," *Science* 134 (10 November 1961), 1525–1526. M. M. Zaret et al., "Laser Photocoagulation of the Eye," *Archives of Ophthalmology* 69 (1963), 97–104.

113. Eric J. Woodbury, paraphrase of an interview, 14 November 1985 (SHL); D. A. Buddenhagen, B. A. Lengyel, F. A. McClung, and G. F. Smith, "An Experimental Laser Ranging System," *IRE International Convention Record* 9, part 5 (1961), 285–290; M. L. Stitch, E. J. Woodbury, and J. H. Morse, "Optical Ranging System Uses Laser Transmitter," *Electronics* 34 (21 April 1961), 51–54.

114. Spiking in ruby lasers had first been spotted in the middle of the summer of 1960 by the team of Bell Laboratory scientists who had replicated Maiman's laser and extended the investigation of its properties. Similar, although less complex, behavior had already been observed and analyzed in microwave ruby masers. See R. J. Collins et al., "Coherence, Narrowing, Directionality, and Relaxation Oscillations in the Light Emission from Ruby," *Physical Review Letters* 5 (1960), 303–305 and C. Kikuchi et al., "Ruby as a Maser Material," *Journal of Applied Physics* 30 (1959),

1061–1067; H. Statz and G. deMars, "Transients and Oscillation Pulses in Masers," in C. H. Townes, ed., *Quantum Electronics* (New York: Columbia University Press, 1960), pp. 530–537.

115. M. L. Stitch, E. J. Woodbury and J. H. Morse, "Repetitive Hair-Trigger Mode of Optical Maser Operation," *Proc. IRE* 49 (1961), 1571–1572.

116. Robert W. Hellwarth, "Theory of the Pulsation of Fluorescent Light from Ruby," *Physical Review Letters* 6 (1961), 9–12.

117. R. W. Hellwarth, "Control of Fluorescent Pulsations," in Singer, ed., *Advances in Quantum Electronics* (cited in note 1), pp. 334–341; Robert W. Hellwarth, interview for the Laser History Project, 28 and 29 May 1985 (SHL).

118. F. J. McClung and R. W. Hellwarth, "Giant Optical Pulsations from Ruby," *Journal of Applied Physics* 33 (1962), 828–829.

119. Seidel, "From Glow to Flow," cited in note 11, and George F. Smith, "The Early Years at Hughes Aircraft Company," *Journal of Quantum Electronics* QE-20 (1984), 582. In the end, the company made use of a mechanical, rotating mirror system for Q-switching first demonstrated by the Army Signal Research and Development Laboratory at Fort Monmouth.

120. E. J. Woodbury, "Raman Laser Action in Organic Liquids," in Grivet and Bloembergen, eds., *Quantum Electronics*, cited in note 1, pp. 1577–1588; E. J. Woodbury, "The Discovery of Stimulated Raman Scattering at Hughes Aircraft," manuscript dated 8 September 1983 (SHL); Woodbury, interview, cited in note 113.

121. E. J. Woodbury and W. K. Ng, "Ruby Laser Operation in the Near IR," *Proc. IEEE* 50 (1962), 2367.

122. Fred C. McClung, paraphrase of an interview, 6 March 1985 (SHL).

123. See their paper "The Raman Laser," in Grivet and Bloembergen, eds., *Quantum Electronics*, cited in note 1, pp. 1589–1596; P. E. Tannenwald and H. J. Zeiger, paraphrase of an interview, 8 January 1986 (SHL).

124. Lawrence Lessing, "The Laser's Dazzling Future," Fortune 67 (June 1963), p. 151.

125. Elliott, "Spotlight," cited in note 15. For the papers, the caveats mentioned in note 5 also apply.

126. Interviews with H. Kogelnik, 2 May 1986, and C. K. N. Patel, 31 May and 4 June 1984 (SHL); paraphrases of interviews with Alan D. White, 30 April 1986, and Joseph E. Geusic, 21 October 1987 (SHL). P. K. Tien to the author, private communication, 7 December 1988.

127. W. R. Bennett jr. "Gaseous Optical Masers," *Applied Optics* supplement (1962), 24–61, here p. 27. D. R. Herriott to J. L. Bromberg, 23 October 1985.

128. G. D. Boyd and J. P. Gordon, "Confocal Multimode Resonator for Millimeter through Optical Wavelength Masers, *Bell System Technical Journal* 40 (1961), 489–508; Gary D. Boyd, paraphrase of an interview for the Laser History Project, 30 September 1986 (SHL).

129. W. R. Bennett jr., deposition, 29 February 1968, Interference #94,835 (courtesy of Patlex Corp.), p. 230.

130. W. W. Rigrod, H. Kogelnik, D. J. Brangaccio, and D. R. Herriott, "Gaseous Optical Maser with External Concave Mirrors," *Journal of Applied Physics* 33 (1962), 743–744; D. J. Brangaccio, "Construction of a Gaseous Optical Maser Using Brewster Angle Windows," *Review of Scientific Instruments* 33 (1962), 921–922; Kogelnik, interview, cited in note 126.

131. Bennett, "Gaseous Optical Masers," cited in note 127, pp. 26, 28.

132. "Proposal To Study the Properties of Laser Devices," pp. 56–57, and Patent Specification, p. 61, R. G. Gould Papers. It was also suggested by Schawlow and Townes in their *Physical Review* paper, p. 1948.

133. Isaac D. Abella, paraphrase of an interview, 22 October 1985 (SHL); Paul Rabinowitz, deposition, 23 May 1968, Interference #94,837 (courtesy of Patlex Corp.).

134. Rabinowitz, deposition, cited in note 133.

135. R. G. Gould, interview, 23 October 1983 (SHL).

136. Abella, paraphrase of an interview, cited in note 133; Herman Z. Cummins, laboratory notebooks, Columbia University Physics Department papers, Box 104, Columbia University Archives.

137. Lawrence Goldmuntz, interview, 21 October 1983 (SHL).

138. Stephen J. Jacobs, memoir, August 1984 (SHL); Bennett, "Gaseous Optical Masers," cited in note 127, pp. 45–46; S. Jacobs, G. Gould, and P. Rabinowitz, "Coherent Light Amplification in Optically Pumped Cs Vapor," *Physical Review Letters* 7 (1961), 415–417; Jacobs, Rabinowitz, and Gould, "Optical Pumping of Cesium Vapor (abstract)," *JOSA* 51 (1961), 477; Rabinowitz, Jacobs, and Gould, "Continuous Optically Pumped Cs Laser," *Applied Optics* 1 (1962), 513–516.

139. Walter L. Faust, memoir, n.d. (ca. May 1983).

140. Bennett, "Gaseous Optical Masers," cited in note 127.

141. Patel, interview, cited in note 126.

142. Ibid.

143. See the articles by Patel, Faust, and McFarlane, and the references thereto, in Grivet and Bloembergen, eds., *Quantum Electronics*, cited in note 1, pp. 507–514, 561–572. See also McFarlane, Faust, Patel, and Garrett, ibid., pp. 573–586.

144. Patel, interview, cited in note 126.

145. Bennett, "Gaseous Optical Masers," cited in note 127, p. 44. Bennett wrote, "It was far from obvious that the lifetimes of the [upper and lower] states were actually suitable for obtaining an inversion and White and Rigden took a rather bold step."

146. A. D. White and J. D. Rigden, "Continuous Gas Maser Operation in the Visible," *Proc. IRE* 50 (1962), 1697.

147. See Bromberg, "Engineering Knowledge in the Laser Field," cited in note 45.

148. White, paraphrase of interview cited in note 3.

149. Ibid.

150. Ernest Braun and Stuart MacDonald, *Revolution in Miniature: The History and Impact of Semiconductor Electronics* (Cambridge: Cambridge University Press, 1978), p. 60.

151. See Russell D. Dupuis, "An Introduction to the Development of the Semiconductor Laser," *IEEE Journal of Quantum Electronics* QE-23 (1987), 651–657, and the references therein. See also Bertolotti, *Masers and Lasers*, cited in note 23, pp. 165–174, and Bela A. Lengyel, *Lasers* (New York: Wiley-Interscience, 2nd ed., 1971), pp. 237–240. Although this paragraph is based on Dupuis's article, I do not deal with John von Neumann's 1953 proposal, because it was made in the different historical context of premaser physics.

152. In a direct semiconductor such as gallium arsenide, an electron in the lowest energy conduction band level has the same momentum as an electron in the highest energy valence band level. In an indirect semiconductor such as silicon, the momenta are different. The excess momentum is usually transferred to the lattice along with a quantum of energy.

153. Benjamin Lax, interview, 15 May 1986 (SHL); Benjamin Lax, "Cyclotron Resonance and Impurity Levels in Semiconductors," in Townes, ed., *Quantum Electronics*, cited in note 114, pp. 428–449; Robert H. Rediker, "Research at Lincoln Laboratory Leading up to the Development of the Injection Laser in 1962," *IEEE Journal of Quantum Electronics* QE-23 (1987), 692–695.

154. J. I. Pankove, "First European Report" (PEM-804, 1/3/57), "Second European Report" (PEM 860, 4/17/57), "Technical Report: Electrically Pumped Semiconductor Lasers" (PEM-2126, 2/2/62) (all courtesy of J. I. Pankove).

155. Nikolai G. Basov, "Semiconductor Lasers," Nobel lecture, 11 December 1964, published in *Science* 149 (20 August 1965), 821–827; Basov, "History of Lasers at Lebedev Physical Institute," cited in note 62 of chapter 2; Dupuis, "Introduction," cited in note 151.

156. A junction is formed by doping one side of a semiconductor crystal in such a way that the mobile charges are positive (p-type) and the other side in such a way that the mobile charges are negative (n-type). When the crystal is attached to electrodes and a current is applied in the proper ("forward") direction, positive charges cross (are "injected") into the n-type region and negative charges are injected into the p-type region. Negative electrons and positive holes now coexist in the same region, the junction, and can produce light by combining ("recombination radiation"). If a sufficiently high current density (or density of injected charges) is maintained, stimulated emission may dominate in the recombination radiation and the diode may pass from spontaneous to stimulated emission. See Jacob Millman, *Microelectronics* (New York: McGraw-Hill, 1979), pp. 26–31, and Lengyel, *Lasers*, cited in note 151, chap. 7.

157. See M. I. Nathan, "Invention of the Injection Laser at IBM," in *IEEE Journal of Quantum Electronics* QE-23 (June 1987), 679–683.

158. R. W. Landauer, interview, 17 October 1984 (SHL); R. W. Landauer, "The History of Work at IBM Research Leading to the Discovery of the Injection Laser," memoir, 5 April 1971, and "Trip Report on the International Conference on Photoconductivity held at Cornell University, August 21–24, 1961," dated 30 August 1961, R. W. Landauer Papers; Gordon Lasher, IBM Notebook 6907, entry of 21 April 1961, courtesy of G. Lasher.

159. Landauer, interview, cited in note 158; "Proposal for a Research and Development Program to find an Electron Injection Laser," 4 January 1962, R. W. Landauer Papers.

160. W. P. Dumke, "Interband Transitions and Maser Action," IBM Research report RC-626, 19 January 1962, published in *Physical Review* 127 (1962), 1559–1563. Dumke had an old interest in semiconductor lasers that had been revived by Landauer's activity. See the conversation with F. H. Dill, W. P. Dumke, F. Stern, and G. Lasher, 18 October 1984 (SHL).

161. GT&E, Semiconductor Material and Devices, Solid State Laboratory, Reports on Project 353 for April–June and September–December 1960; "Summary of work at General Telephone & Electronics Laboratories. . . January 1961 through December 1964," compiled by J. F. Black with S. Mayburg, S. M. Ku, and H. Lockwood (SHL). Lockwood made this discovery while studying the current-voltage characteristics of gallium arsenide diodes, while Ku was investigating low-temperature varactor diodes.

162. "Summary," cited in note 161. General Telephone and Electronics Laboratories, "Proposal for Electron Injection Laser," 5 January 1962, J. F. Black Papers. GT&E ultimately got a contract for this work from Fort Belvoir (S. Mayburg, private communication).

163. J. I. Pankove, private communication, 5 December 1988.

164. J. I. Pankove to H. Johnson, 21 January 21, 1963, and W. M. Webster to J. I. Pankove, 24 January 1963 (both courtesy of J. I. Pankove).

165. J. I. Pankove and M. J. Massoulié, "Injection Luminescence from Gallium Arsenide," *Bulletin of the American Physical Society* Series II 7 (1962), 88.

166. S. Mayburg to J. H. Crawford jr., 12 July 1962 (courtesy of J. F. Black). In the event, the GT&E paper was turned down by *Physical Review Letters* because of Pankove and Massoulié's prior announcement, and Mayburg reported the results at the March 1962 meeting of the American Physical Society.

167. R. W. Landauer, private communication; Nathan, "Invention," cited in note 157.

168. R. W. Keyes, "The Current State of Injection Laser Research," 5 June 1962 (courtesy of R. W. Landauer) (SHL). Conversation with Dill et al., cited in note 160.

169. Nathan, interview, cited in note 4.

170. Ibid.

171. Rediker, "Research at Lincoln Laboratory," cited in note 153, quotation on p. 692; Robert H. Rediker, Ivars Melngailis, and Aram Mooradian, "Lasers, Their Development, and Applications at M.I.T. Lincoln Laboratory," *IEEE Journal of Quantum Electronics* QE-20 (1984), 603, and references cited therein. Quarterly Progress Reports, Solid State Research, for 15 January, 15 April, and 15 July 1961 (MIT Lincoln Laboratory Library Archives).

172. *New York Times*, 10 July 1962, p. 23.

173. Nathan, interview, cited in note 4.

174. Robert. N. Hall, "Injection Lasers," *IEEE Transactions on Electron Devices* ED-23 (July 1976), 700–704, here p. 700.

175. Lengyel, *Lasers*, cited in note 151, p. 238. Lengyel points out that Basov's group independently derived these conditions at the same time.

176. R. N. Hall, interview by Jeff Hecht, 11 October 1984 (SHL). See *Lasers & Applications, Laser Pioneer Interviews* (Torrance, California: High Tech Publications, 1985), p. 133.

177. Hall, "Injection Lasers," cited in note 174.

178. Ibid., p. 704.

179. Hall, interview, cited in note 176.

180. Hall, "Injection Lasers," cited in note 174, p. 701.

181. N. Holonyak jr., "Semiconductor Alloy Lasers—1962," *IEEE Journal of Quantum Electronics* QE-23 (1987), 684–691; N. Holonyak jr., private communication, February 1988.

182. The story is given in detail in Rediker, "Research at Lincoln Laboratory," cited in note 153. Lax, in the interview cited in note 153, recalled, however, that he had directed Keyes and Rediker to work on a semiconductor laser as early as late 1960 or early 1961.

183. Nathan, "Invention," cited in note 157.

184. Nathan, interview, cited in note 4.

185. S. Mayburg, J. F. Black, S-M Ku, and H. Lockwood, private communication.

186. Braun and MacDonald, *Revolution in Miniature*, cited in note 150, chapters 9 and 10.

187. Hall, "Injection Lasers," cited in note 174, p. 701.

188. Holonyak, private communication, cited in note 181.

189. *Electronics* 35 (23 November 1962), 14–15. The author, Michael F. Wolff, concluded that it wouldn't.

190. Gerald W. Brock, *The U.S. Computer Industry: A Study of Market Power* (Cambridge: Ballinger, 1975), pp. 16, 23, 24.

191. John A. Armstrong, interview, 17 October 1984 (SHL); Robert Saltonstall jr. et al., *The Commercial Development and Application of Laser Technology* (New York: Hobbs, Dorman & Co., 1965).

192. Oskar A. Reimann, "On All-Optical Computer Techniques," in James T. Tippett et al., eds., *Optical and Electro-Optical Information Processing* (Cambridge: MIT Press, 1965), pp. 247–252; Walter F. Kosonocky, "Laser Digital Devices," ibid., pp. 269–304.

193. *Time* (2 January 1961), 40ff. *Time* and other magazines display a strong tendency to cluster new scientific events around a small number of easily recognizable names. Townes is universally named, beginning with the earliest articles. By 1962, Maiman's name is added. Schawlow is almost never given a role as coauthor of the pioneer theoretical article, although he is quoted on other matters. See Herbert J. Gans, *Deciding What's News* (New York: Pantheon, 1979).

194. *U.S. News* 52 (2 April 1962), 47–50, and 54 (25 February 1963), 80–82; *Life* 54 (11 January 1963), 46–50; *Reader's Digest* 82 (February 1963), 97–100; *New York Times Magazine* (8 September 1963), 27ff.; *Saturday Evening Post* 237 (24 October 1964), 68–71.

195. Thus an article in *Time* (17 October 1960) based on an interview with A. L. Schawlow mentions communications, laser interferometry, and the use of lasers for catalysis of chemical reactions.

196. L. S. Smullin and G. Fiocco, "Project Luna See," *Proc. IRE* 50 (1962), 1703–1704; *New York Times*, 11 May 1962, 1:5; Lessing, "The Laser's Dazzling Future," cited in note 124.

Chapter 5: Out of the Laboratory and into the Marketplace

1. Robert W. Seidel, "From Glow to Flow: A History of Military Laser Research and Development," *Historical Studies in the Physical and Biological Sciences* 18(1) (1987), 111–147, here p. 134.

2. L. R. Bittman, "Limitations on Lasers for Deep Space Communications," *IEEE Transactions on Communications and Electronics* 83 (March 1964), 170–173, was one of several who put forth the thought, though not yet the slogan, in late 1962.

3. Robert D. Haun et al., "The Application of Lasers to Industry," *IEEE Transactions on Industry and General Applications* IGA-4 (1968), 379.

4. From 1961 through 1964, the figures represent a count of the number of abstracts for each year, from the subject index for 1961–1964. For the period 1965–1968, nine pages out of a total of 30+ pages devoted to lasers were sampled in the 1965–1968 index.

5. Gary K. Klauminzer, "Twenty Years of Commercial Lasers—A Capsule History," *Laser Focus/Electro-Optics* (December 1984), 54ff.

6. Stephen E. Harris et al., "Controlling Laser Oscillation," *Electronics* 38 (20 September 1965), 101–105. These techniques were for axial modes, but transverse modes were also controlled.

7. L. E. Hargrove et al., "Locking of He-Ne Laser Modes . . . ," *Applied Physics Letters* 5 (1964), 4–5; Harris et al., "Controlling Laser Oscillations," cited in note 6; A. J. DeMaria et al., "Ultrashort Light Pulses," *Science* 156 (23 June 1967), 1557–1568; Peter W. Smith, "Mode-Locking of Lasers," *Proc. IEEE* 58 (September 1970), 1342–1357.

8. National Science Foundation, "National Patterns of R&D Resources, Funds & Personnel in the United States, 1953–1978/79," NSF publication 78–313.

9. Fred P. Burns, paraphrase of an interview, 30 March 1987 (SHL).

10. "Marketers' and Buyers' Guide," *Laser Focus* 2 (15 January 1966), 3; see also *Barron's* (28 February 1966), 3ff.

11. Electronic Industries Association, *Electronic Market Data Book*, 1970, p. 63, and 1971, p. 67; "Review and Outlook, 1971," *Laser Focus* 7 (January 1971), 24ff.; all figures are in current dollars.

12. Laser Institute of America (formerly Laser Industry Association) Board of Directors, Minutes, 1967–1971 (courtesy of LIA Headquarters); Malcolm L. Stitch, "Evolution of a Trade Association," *Space Age News* (October 1968), 9–14; Arthur Lubin, paraphrase of an interview, 17 June 1985 (SHL); file, "Laser Focus, 1967–1974," W. B. Bridges Papers; *Electronics* 40 (16 October 1967), 26.

13. *Missiles and Rockets* (1 June 1964), 26; *Electronics* 37 (19 October 1964), 96.

14. These numbers are rough. Ambiguous entries were generally simply discarded, and no attempt was made to correct for the fact that there were greater delays in indexing articles from non-English-speaking nations than from English-speaking ones.

15. French scientists G. Convert, M. Armand, and P. Martinot-LaGarde found lasing in mercury ions independently, but they published later. See "Effet laser dans les mélanges mercure-gaz rares," *Comptes Rendus* 258 (23 March 1964), 3259–3260, and M. Bertolotti, "Twenty-Five Years of the Laser: The European Contribution to its Development," *Optica Acta* 32 (1985), 962.

16. Letter from Earl Bell to J. L. Bromberg, 10 June 1987 (SHL); see also Arnold L. Bloom to J. L. Bromberg, 18 May 1987 (SHL). W. E. Bell, "Visible Laser Transitions in Hg+," *Applied Physics Letters* 4 (15 January 1964), 34–35.

17. *Aviation Week & Space Technology* 78 (22 April 1963), 54ff; Seidel, "From Glow to Flow," cited in note 1, pp. 128–129; *Electronics* 36 (22 February 1963), 30.

18. *Electronics* 37 (24 August 1964), 28.

19. A. L. Bloom, "Gas Lasers," *Applied Optics* 5 (1966), 1502–1503.

20. William B. Bennett jr., interview, 27 October 1987 (SHL).

21. J. D. Rigden, private communication, 28 October 1987.

22. Under the leadership of Donald C. Forster, the group had in fact already been inching its way from microwave tubes toward laser research; see W. B. Bridges, interview by Richard Cunningham, in *Lasers & Applications* (September 1985), 83–88.

23. Hughes Research Laboratories Quarterly Reports for 1963, Project 3232, and W. B. Bridges and D. F. Hotz, "Gas Lasers for Optical Receivers and Transmitters," typescript, November 1983, W. B. Bridges Papers.

24. Bridges, interview, cited in note 22; also William B. Bridges, "Ion Lasers: A Retrospective," in Carl B. Collins, ed., *Proceedings of the International Conference on Lasers, '81*, 14–18 December 1981, pp. 1–2. Helium's role in gas lasers was a general problem at this time; see R. A. Panaanen et al., "Laser Action in Cl_2 and He-Cl_2," *Applied Physics Letters* 3 (1963), 154–155, and J. D. Rigden and A. D. White, "Optical Maser Action in Iodine," *Nature* 198 (1963), 774.

25. Bridges, interview, cited in note 22, p. 84.

26. Hughes Aircraft Company Technical Journal No. C1239 (assigned to William B. Bridges), p. 63, W. B. Bridges Papers; Hughes Research Laboratories, Quarterly Report for Project 3232: Gaseous Lasers, 1 January–31 March 1964.

27. Donald C. Forster to Kenneth Hutchinson, Aeronautical Systems Division, Wright Patterson Air Force Base, 28 February 1964, W. B. Bridges Papers.

28. Both were published later. See Bennett et al., *Applied Physics Letters* 4 (1964), 180–182, and Convert et al., *Comptes Rendus* 258 (4 May 1964), 4467–4469. See also Bennett, interview, cited in note 20, and P. Laurès, "Etat actuel des recherches et des applications dans le domaine des lasers à gaz," *Revue générale de l'électricité* 76 (1967), 1225–1244.

29. E. I. Gordon, interview, 3 June 1984 (SHL).

30. *Hughesnews* 23 (22 May 1964), Hughes Aircraft Company Archives.

31. Gordon, interview, cited in note 30; E. I. Gordon, E. F. Labuda, and W. B. Bridges, "Continuous Visible Laser Action in Singly Ionized Argon, Krypton, and Xenon," *Applied Physics Letters* 4 (15 May 1964), 178–180; Joan Lisa Bromberg, "William B. Bridges and the Ion Laser," *Laser Topics* 8 (fall 1986), 2.

32. Hughes Research Laboratories, Quarterly Report for Project 3232: Gaseous Lasers, 1 April–30 June 1964, W. B. Bridges Papers.

33. This was the upper limit for practical lasers, although experimental models had been driven higher: Arnold L. Bloom, "Gas Lasers," *Applied Optics* 5 (October 1966), 1504.

34. *Electronics* 37 (28 December 1964), 17; *Raytheon News* 14 (January 1965), 4 (courtesy of Raytheon Company's Public Relations Office); Monthly Progress Reports, Research Division, Theoretical Physics Group, 1959–December 1966, Raytheon Research Division Library; quotation from Memo M1708, 18 November 1964.

35. Roy A. Paananen, "Progress in Ionized-Argon Lasers," *IEEE Spectrum* 3 (June 1966), 99; *Hughesnews* 23 (22 May 1964), Hughes Aircraft Company Archives.

36. Barry Miller, "Air Reconnaissance Aided by Line-Scanning Laser Camera," *Aviation Week & Space Technology* 82 (26 April 1965), 80ff., and "Air Force Using Line-Scan Camera," ibid. 92 (26 January 1970), 51ff.; W. B. Bridges, interview, 28 January 1985 (SHL). Although a few units were ultimately delivered to Vietnam, argon reconnaissance was abandoned in the early 1970s in favor of an infrared technology; see A. S. Halsted, paraphrase of an interview, 1 December 1987 (SHL).

37. Francis L'Esperance jr., interview by Jeff Hecht, *Lasers & Applications* (May 1986), 79ff. Francis A. L'Esperance jr.,"An Ophthalmic Argon Laser Photocoagulation System . . . ," *Transactions of the American Ophthalmological Society* 66 (1968), 827–904.

38. By 1969, they had 30–40 employees. At this point, short of capital and unable to raise funds in the depressed security market of late 1968–1969, Orlando Research sold itself to Ivashuk Manufacturing Corp. and became Control Laser Corporation. See John Tracy, paraphrase of an interview, 17 December 1987 (SHL), and the prospectus for Control Laser Corporation, 2 October 1978.

39. Klauminzer, "Twenty Years," cited in note 5, p. 58. Coherent Radiation, Inc., is discussed below. *Hughesnews* (2 December 1966), Hughes Aircraft Company Archives; A. Stevens Halsted, paraphrase of an interview, 1 December 1987 (SHL). Earlier, in 1965, the Research Laboratories produced small quantities for sale (W. B. Bridges, private communication, 21 March 1989).

40. J. E. Geusic, W. B. Bridges, and J. I. Pankove, "Coherent Optical Sources for Communication," *Proc. IEEE* 58 (October 1970), 1419–1439, esp. 1422–1426; Peter O. Clark, "Advances in Gas Lasers," *IEEE International Convention Record* 15, part 7 (1967), 69–79; D. C. Forster, "CW Argon Laser Development," memorandum, 14 October 1964, W. B. Bridges Papers.

41. Klauminzer, "Twenty Years," cited in note 5, reports that a firm called Lexel spun off from Coherent to pursue beryllium-oxide tubes, after Coherent chose to develop graphite.

42. See the papers by Geusic et al. and by Clark cited in note 40.

43. C. K. N. Patel, interview, 31 May and 4 June 1984 (SHL).

44. Arthur Erikson, "What's New in Lasers?," *Electronics* 36 (1 March 1963), 14–16 (a report on the Paris conference).

45. For further details, see C. K. N. Patel, "High-Power Carbon Dioxide Lasers," *Scientific American* 219 (August 1968), 22–33.

46. Patel, interview, cited in note 43; Notebook #5 (38083), C. K. N. Patel Papers. (I am grateful to Dr. Patel for making these notebooks available to me.)

47. L. E. S. Mathias and J. T. Parker, "Stimulated Emission in the Band Spectrum of Nitrogen," *Applied Physics Letters* 3 (1 July 1963), 16–18, and "Visible Laser Oscillations from Carbon Monoxide," *Physics Letters* 7 (1963), 194–196.

48. C. K. N. Patel, W. L. Faust, and R. A. McFarlane, "Laser Action on Rotational Transitions of the Σ_u^+-Σ_g^+ Vibrational Band of CO_2." *Bulletin of the American Physical Society* 9 (1964), 500; Patel, Notebook #8 (39595), C. K. N. Patel Papers.

49. J. E. Morgan and H. I. Schiff, "The Study of Vibrationally Excited N_2 Molecules with the Aid of an Isothermal Calorimeter," *Canadian Journal of Chemistry* 41 (1963), 903–912, and references therein.

50. C. K. N. Patel. "Selective Excitation Through Vibrational Energy Transfer and Optical Maser Action in N_2-CO_2," *Physical Review Letters* 13 (1964), 617–619.

51. For references to the papers of the Legays and their collaborators, see Bertolotti, "Twenty-Five Years," cited in note 15, and Laurès, "L'Etat actuel," cited in note 28.

52. J. C. Polanyi, "Proposal for an Infrared Maser Dependent on Vibrational Excitation," *Journal of Chemical Physics* 34 (1961), 347–348 (based on a talk given 8 June 1960); see also the letter dated 2 January 1984 by John C. Polanyi (SHL).

53. S. Millman, ed., *A History of Engineering and Science in the Bell System: Physical Sciences (1925–1980)* (AT&T Bell Laboratories, 1983), pp. 168–172, and the references therein. This source reports CW power over 100 kilowatts by 1979 and pulsed powers more than a terawatt.

54. C. K. N. Patel, "CW High Power N^2-CO^2 Laser," *Applied Physics Letters* 7 (1965), 15–17; C. K. N. Patel, P. K. Tien, and J. H. McFee, "CW High-Power CO_2-N_2-He Laser," ibid., 290–292; Patel, Notebooks #9 (40485), #10 (40060), #11 (42734). G. Moeller and J. D. Rigden at Perkin-Elmer used CO_2-N_2-He independently and published an issue earlier than Patel's group: "High-Power Laser Action in CO_2-He Mixtures," ibid., 274–276.

55. Patel, interview, cited in note 43.

56. Bloom, "Gas Lasers," cited in note 19, 1501, 1510, 1512. Q-switching was demonstrated by M. A. Kovacs, G. W. Flynn, and A. Javan, using both CO_2 and N_2O; see "Q-Switching of Molecular Laser Transitions," *Applied Physics Letters* 8 (1966), 62–63.

57. Seidel, "From Glow to Flow," cited in note 1; R. W. Seidel, private communication, 30 November 1988.

58. Later, however, a sharp competition developed between the two companies over CO_2 lasers.

59. The company did not turn a profit, however, until it supplemented its carbon dioxide lasers with a line of ion lasers. Coherent won the argon ion market with its

second model, introduced at the end of 1968, which cut weight by more than 90% and price by 50%, while retaining two-thirds of the power output. *Electronic News* (14 October 1968), 38 (courtesy of David Belforte); James L. Hobart, interview, 20 January 1984, untranscribed tape, Niels Bohr Library, American Institute of Physics; H. Gauthier, R. Rorden, and E. DeRousse, private communication, 20 January 1984; Prospectus, 19 May 1970, and Report to the Stockholders, January 1971 (both courtesy of J. L. Hobart).

60. The names of the departments and the responsibilities of Bell Laboratories staff members changed from year to year. In this paragraph, I use the BTL Organizational Directory for January 1965, AT&T Archives.

61. J. E. Geusic and H. E. D. Scovil, "Microwave and Optical Masers," *Reports on Progress in Physics* 27 (1964); the analysis is on pp. 288–301.

62. J. E. Geusic, paraphrase of an interview, 21 October 1987 (SHL).

63. For a survey of other solid-state lasers to 1966, see Z. J. Kiss and R. J. Pressley, "Crystalline Solid Lasers," *Applied Optics* 5 (1966), 1474–1486.

64. Geusic, paraphrase of an interview, cited in note 62; Legrand G. Van Uitert, memoir, 7 June 1983 (SHL); J. E. Geusic, H. M. Marcos, and L. G. Van Uitert, "Laser Oscillations in Nd-doped Yttrium Aluminum, Yttrium Gallium and Gadolinium Garnets," *Applied Physics Letters* 4 (April 1964), 182–184; J. E. Geusic, M. L. Hensel, and R. G. Smith, "A Repetitively Q-switched, Continuously Pumped YAG:Nd Laser," *Applied Physics Letters* 6 (May 1965), 175–177; M. DiDomenico jr., J. E. Geusic, H. M. Marcos, and R. G. Smith, "Generation of Ultrashort Optical Pulses by Mode-Locking the YAlG:Nd Laser," *Applied Physics Letters* 8 (April 1966), 180–183; Millman, ed., *History,* cited in note 53, p. 172.

65. Richard T. Daly, interview by Paul Forman, 28 May 1984, untranscribed tape, Niels Bohr Library, American Institute of Physics; *Aviation Week & Space Technology* 87 (21 August 1967), 95.

66. R. J. Pressley, paraphrase of an interview, 30 May 1985 (SHL); Holobeam Securities and Exchange Commission form 10K, for 30 September 1974.

67. Klauminzer, "Twenty Years," cited in note 5, pp. 74–75.

68. Geusic, Bridges, and Pankove, "Coherent Optical Sources," cited in note 40, 1431–1435; Walter Koechner, "YAG Challenges Carbon Dioxide in High C-W Power," *Laser Focus* 5 (September 1969), 29–34.

69. Geusic, paraphrase of an interview, cited in note 62.

70. Thomas P. Hughes, "The Evolution of Large Systems," in Wiebe E. Bijker, Thomas P. Hughes, and Trevor J. Pinch, eds., *The Social Construction of Technological Systems* (Cambridge: MIT Press, 1987), pp. 51–82.

71. See p. 1431 of the article by Geusic, Bridges, and Pankove cited in note 40 and p. 29 of the article by Koechner cited in note 68; *Laser Focus* 6 (January 1970), 29; Klauminzer, "Twenty Years," cited in note 5, p. 74–75.

72. As well as being visible in organization charts, the division is reflected in the interviews of the Sources for Laser History. See, for example, those of Alan D. White and Eugene I. Gordon. For Hughes Aircraft Company, see the interview of William B. Bridges.

73. Robert C. Rempel, interview, 18 January 1984, untranscribed tape, Niels Bohr Library, American Institute of Physics.

74. Thomas Maguire, "Laser Welds Copper Leads," *Electronics* 36 (25 October 1963), 88ff.

75. The pulse repetition rate was restricted by the heating of the laser. The more efficiently pump energy could be translated into laser beam energy, the smaller the residual energy that would be transformed into heat.

76. *Electronics* 37 (24 August 1964), 28.

77. Plans for such symposia, sponsored jointly by federal agencies and the Electronics Industries Association, are described in M. L. Stitch, "Evolution of a Trade Association," *Space Age News* (October 1968), 9–13. See also *Electronics* 33 (29 July 1960), 46.

78. This interpretation is disputed by William B. Bridges, who writes, "I think your division of 'physicists doing research' and 'engineers doing systems' is artificial. . . . I believe most device people were fully aware of the laser properties that were desirable, perhaps even more than 'systems' people, who had not thought in terms of *optical* systems before. I think this tone creates a wrong view of how events developed" (letter to J. L. Bromberg, 21 March 1989).

79. See Kurt E. Shuler, "Editorial Foreword," *Applied Optics*, supplement 2 (*Chemical Lasers*) (1965), 1–2; Shuler makes the territorial claim half-jocularly. One laser using complex molecules that was much studied in the mid-1960s, and that was particularly interesting to the chemists, was the chelate laser. A chelate is a molecule in which a central rare earth or metal atom is bonded to a number of organic groups, which surround it like a cage. See Alexander Lempicki and Harold Samuelson, "Liquid Lasers," *Scientific American* 216 (June 1967), 80–90.

80. Polanyi, "Proposal," cited in note 52.

81. Kenneth C. Herr and George C. Pimentel, "A Rapid-Scan Infrared Spectrometer . . . ," *Applied Optics* 4 (1965), 25–30.

82. G. C. Pimentel, "Chemical Lasers," Scientific American 214 (April 1966), 32–39; George C. Pimentel, interview, 17–18 May 1984 (SHL). G. C. Pimentel, graduate log book, G. C. Pimentel Papers.

83. J. V. V. Kasper and G. C. Pimentel, "Atomic Iodine Photodissociation Laser," *Applied Physics Letters* 5 (1965), 231–233; D. Porret and C. F. Goodeve, "The Continuous Absorption Spectra of Alkyl Iodides . . . ," *Proc. Royal Society of London* A165 (1938), 31–49; Pimentel, interview, cited in note 82.

84. Raytheon, Special Microwave Devices Operation, "Catalog of Microwave Devices, Materials and Lasers," ca. 1962 (courtesy of C. Bowness).

85. "Chemical Pumping Boosts Laser Power," *Chemical Week* (14 March 1964), 73.

86. Kurt E. Shuler, private communication, 19 July 1988.

87. The *Applied Optics* supplement 2 is the record of this meeting. See Shuler's foreword, cited in note 79. In its purpose, the conference resembles the coherence conference of June 1960, discussed in the section on professional societies in chapter 4. See also "Chemical Laser Efforts Broaden Scope of Laser Research," *Chemical & Engineering News* (8 February 1965), 38–40. Reference to an earlier

Soviet conference is made in Mario Bertolotti, *Masers and Lasers: An Historical Approach* (Bristol: Adam Hilger, 1983), p. 163.

88. Pimentel, "Chemical Lasers," cited in note 82.

89. See also S. S. Penner, "On Iraser Detectors for Radiation Emitted from Diatomic Gases and Coherent Infrared Sources," *Journal of Quantitative Spectroscopy and Radiative Transfer* 1 (1961), 163–168.

90. Arthur N. Chester, "Chemical Lasers," in E. R. Pike, ed., *High-Power Gas Lasers, 1975* (London: The Institute of Physics, 1976), p. 162.

91. Dyes are organic compounds that absorb in the visible. See F. P. Schaefer, "Principles of Dye Laser Operation," in F. P. Schaefer, ed., *Dye Lasers* (New York: Springer Verlag, 1973), pp. 6–7.

92. The dye laser was also discovered independently by Mary L. Spaeth and D. P. Bortfield at Hughes and by Fritz P. Schaefer and coworkers in Germany. Both of these groups published later. See Bertolotti, *Masers and Lasers*, cited in note 87, pp. 160–163.

93. The term *tuning* was also used during the 1960s to refer to a process of selecting out one from among a number of the discrete wavelengths at which a laser could oscillate. I shall not use it in this sense.

94. Arthur L. Schawlow, "Spectroscopy in a New Light," Nobel Lecture delivered 8 December 1981, *Reviews of Modern Physics* 54(3) (July 1982), 697–707.

95. For some of these methods, see Ali Javan, "Modern Methods in Precision Spectroscopy: A Decade of Developments," in M. S. Feld, A. Javan, and N. A. Kurnit, eds., *Fundamental and Applied Laser Physics* (New York: Wiley 1973), pp. 295–334.

96. T. W. Hansch, interview, 19 January 1984 (SHL).

97. S. Akhmanov and R. V. Khochlov, "Parametric Amplifiers and Generators of Light," *Soviet Physics. Uspekhi* 9 (September–October 1966), 210–222, quotation on p. 210. This article also gives references to Soviet work, in this respect supplementing the review article of S. E. Harris, "Tunable Optical Parametric Oscillators," *Proc. IEEE* 57 (1969), 2096–2113.

98. See the references in the two articles cited in note 97.

99. The frequencies w_1 and w_2 are governed by the two simultaneous equations $w_1 + w_2 = w_{pumping}$ and $w_1 n(w_1) + w_2 n(w_1) = w_p n(w_p)$, where $n(w)$ is the index of refraction, which is a function of frequency. See Harris, "Tunable Optical Parametric Oscillators," cited in note 97.

100. At the same time, the Soviets achieved the effect in potassium dihydrogen phosphate. J. Giordmaine and R. C. Miller, "Tunable Coherent Parametric Oscillation in $LiNbO_3$ at Optical Frequencies," *Physical Review Letters* 14 (June 1965), 973–976; S. A. Akhmanov et al., "Observation of Two-Dimensional Parametric Interaction of Light Waves," *Soviet Physics, JETP Letters* 2 (1965), 285–288. See also J. A. Giordmaine, interview, 31 May 1984 (SHL).

101. Dietrich Meyerhofer and Rubin Braunstein, "Frequency Tuning of GaAs Laser Diode by Uniaxial Stress," *Applied Physics Letters* 3 (November 1963), 171–172. Subsequently, diode lasers were more effectively tuned by varying the current.

102. H. E. Puthoff et al., "Tunability of the Raman Laser," *Journal of Applied Physics* 37 (1966), 860–864.

103. L. F. Johnson, H. J. Guggenheim, R. A. Thomas, "Phonon-Terminated Optical Masers," *Physical Review* 149 (1966), 179–185.

104. P. P. Sorokin, J. J. Luzzi, J. R. Lankard, and G. D. Pettit, "Ruby Laser Q-Switching Elements Using Phthalocyanine Molecules in Solution," *IBM Journal of Research and Development* 8 (1964), 182–184.

105. P. P. Sorokin, "Contributions of IBM to Laser Science—1960 to the Present," *IBM Journal of Research and Development* 23 (1979), 480.

106. Schaefer, "Principles of Dye Laser Operation," cited in note 91, p. 2.

107. Adapted from Schaefer, ed., *Dye Lasers*, cited in note 91, p. 34.

108. F. P. Schaefer, W. Schmidt, and J. Volze, "Organic Dye Solution Laser," *Applied Physics Letters* 9 (1966), 306–309; Bela A. Lengyel, *Lasers* (New York: Wiley-Interscience, 2nd ed., 1971).

109. Peter P. Sorokin, interview by Jeff Hecht, 27 September 1984 (SHL).

110. *Laser Focus* 3 (June 1967), 12ff. Flash lamps made the whole spectrum accessible because their frequencies went into the ultraviolet. They had several disadvantages; the major one of overheating the dye solution was later solved by flowing the dye through the laser tube.

111. B. H. Soffer and B. B. McFarland, "Continuously Tunable, Narrow-Band Organic Dye Lasers," *Applied Physics Letters* 10 (1967), 266–267. The tuning was accomplished by rotating the grating. The grating served as an extremely wavelength-selective mirror. See the next section for Hansch's subsequent reduction of the width to 0.004 Å.

112. O. G. Peterson, S. A. Tuccio, and B. B. Snavely, "CW Operation of an Organic Dye Solution Laser," *Applied Physics Letters* 17 (1970), 245–247.

113. Theodor W. Hansch, "Applications of Dye Lasers," in Schaefer, ed., *Dye Lasers*, cited in note 91, pp. 194–270 (see pp. 194–197).

114. Schaefer, "Principles of Dye Laser Operation," cited in note 91, p. 1.

115. R. E. Slusher, C. K. N. Patel, and P. A. Fleury, "Inelastic Light Scattering from Landau Level Electrons in Semiconductors," *Physical Review Letters* 18 (1967), 77–79; C. K. N. Patel and R. E. Slusher, "Light Scattering by Plasmons and Landau Levels of Electron Gas in InAs," *Physical Review* 167 (1968), 413–415; C. K. N. Patel and E. D. Shaw, "Tunable Stimulated Raman Scattering from Conduction Electrons in InSb," *Physical Review Letters* 24 (1970), 451–455.

116. Patel, interview, cited in note 43.

117. W. B. Bridges jr., "Introduction," *Applied Optics* 11 (February 1972) 231.

118. Radar is also well represented, but it is a technology with such diverse uses, from weaponry to meteorology to surveying, that I have not chosen to list it as a distinct application.

119. Frontrunners commanded more then 90 entries. The runners up had more than 25. The *Abstracts* do not give an entirely faithful picture of the distribution of research effort because they exclude classified publications, and underrepresent applications in biology and medicine. A distortion is also introduced by the fact that different companies had different policies with respect to publication. Note, however, that the *Abstracts* also cannot be interpreted as giving a portrait of only the

purely civilian effort. For one thing, unclassified bits and pieces of weapons programs could be, and were, published in the open literature. For another, civilian and military applications can only be separated with a finer net than here used. The measurement of the velocity distribution in fluid flow is useful to the designers of both military aircraft and automobile engines. Earth strain is measured to monitor both earthquakes and nuclear bomb tests. Research on laser damage to the retina underlies both ophthalmic surgery and the development of weapons to blind enemy troops.

120. These advantages are rehearsed in a large number of semipopular articles from the 1960s. See, for example, Clarence F. Luck jr., "Lasers for Communication," *Electronic Progress* (a Raytheon house journal) VIII (winter 1964), 37–40, or Stewart E. Miller, "Communication by Laser," *Scientific American* 214 (January 1966), 19–27.

121. Bernard Cooper, "Optical Communications in the Earth's Atmosphere," *IEEE Spectrum* 3 (July 1966), 83–92.

122. *Electronics* 36 (18 October 1963), 24; 39 (31 October 1966), 46, 48; *Aviation Week & Space Technology* 91 (10 November 1969), 75ff.; 92 (29 June 1970), 42; *Proc. IEEE* 53 (December 1965), 2140–2141; *IEEE Transactions on Broadcasting* BC-10 (February 1964), 4–7; James Vollmer, "Applied Lasers," *IEEE Spectrum* 4 (June 1967), 66–70. Frank E. Goodwin, "A Review of Operational Laser Communication Systems," *Proc. IEEE* 58 (1970), 1746–1752. On degradation processes, see H. Kressel and N. E. Byer, "Physical Basis of Noncatastrophic Degradation in GaAs Injection Lasers," *Proc. IEEE* 57 (1969), 25–33. See also the epilogue to this book.

123. R. Smelt, "Some Aspects of the Space Communication Problem," *Canadian Aeronautical Journal* 7 (June 1961), 235–241.

124. Monte Ross, "The History of Space Laser Communications," *S.P.I.E. 885*, Proceedings of the O-E LASE Conference, January 1988, Section on Free Space Laser Communication Technologies, pp. 2–9; Earl J. Reinbolt and Joseph L. Randall, "How Good are Lasers for Deep-Space Communications?," *Astronautics and Aeronautics* 5 (April 1967), 64–70. J. I. Bowen et al., "Report of a Trip to Marshall Space Flight Center—Case 28503," 12 June 1967, R. Kompfner Papers, Box 3, AT&T Archives.

125. See Bittman, "Limitations on Lasers," cited in note 2.

126. Donald F. Nelson, "The Modulation of Laser Light," *Scientific American* 218 (June 1968), 17–23.

127. Ross, "History," cited in note 124.

128. Nilo Lindgren, "Optical Communications—A Decade of Preparations," *Proc. IEEE* 58 (October 1970), 1410–1418.

129. Lindgren, on p. 1415 of the article cited in note 128, estimated nearly 100 people worked on R&D relevant to optical communication at Bell Laboratories in 1970.

130. See, inter alia, R. Kompfner, "Thoughts on Optical Maser Research, Development and Publication Policy . . . " 12 March 1963 (Box 4), and "Comments on 'Light in Metropolitan Interoffice Trunks' . . . ," 19 January 1971 (Box 1), R. Kompfner Papers, AT&T Archives.

131. See, for example, Luck, "Lasers for Communication," cited in note 120, p. 38.

132. Miller, "Communication by Laser," cited in note 120; R. Kompfner, "Optical Communications," *Science* 150 (8 October 1965), 149–155.

133. S. Millman, ed., *A History of Engineering and Science in the Bell System: Communications Sciences (1925–1980)*, (AT&T Bell Laboratories, 1984), pp. 282–283.

134. James Martin, *Future Developments in Telecommunications* (Englewood Cliffs, New Jersey: Prentice-Hall, 1971), pp. 35–49.

135. S. E. Miller and L. C. Tillotson, "Optical Transmission Research," *Applied Optics* 5 (October 1966), 1538–1548.

136. A. J. DeMaria et al., "Ultrashort Light Pulses," *Science* 156 (23 June 1967), 1557–1568; Richard T. Denton, "The Laser and PCM," *Bell Laboratories Record* 46 (June 1968), 175–179.

137. Luck, "Lasers for Communications," cited in note 120, p. 38; Cooper, "Optical Communications," cited in note 121, p. 83; Vollmer, "Applied Lasers," cited in note 122, p. 66.

138. Sami Faltas, "The Invention of Fibre-Optic Communications," *History and Technology* 5 (1988), 31–49.

139. S. E. Miller to E. I. Gordon, "Injection Laser Diodes from Optical Transmission," case 22098, 2 December 1970, and "Optical Fibers as a Transmission Medium" by S. E. Miller, included in L. C. Tillotson to R. Kompfner, "Accomplishments During 1970, Lab 124," 20 October 1970 (both in Box 2); R. Kompfner to W. E. Danielson, "Comments on 'Light in Metropolitan Office Trunks,'" 19 January 1971 (Box 1), all in R. Kompfner Papers, AT&T Archives.

140. Henry Kressel, "The Small, Economy-Size Laser," *Laser Focus* (November 1970), 45–49; D. H. Newan and S. Ritchie, "Sources and Detectors for Optical Fibre Communications Applications: The First 20 Years," *Proc. IEE* 133J (June 1986), 213–228.

141. See the epilogue; also Morton B. Panish and Izuo Hayashi, "A New Class of Diode Lasers," *Scientific American* 225 (July 1971), 32–40.

142. C. H. Townes, "Proposal for Research on a Maser to Amplify or Oscillate at Infrared Frequencies," 9 July 1958, papers of the Columbia University Physics Department, Columbia University Archives; G. F. Smith to A. V. Haeff, "Laser Applications," 21 July 1960, Hughes Aircraft Company Interdepartmental Memo (courtesy of G. F. Smith). These are meant as examples, not as the loci of the first mention of these applications.

143. Klaus D. Mielenz, "Length Measurement and Laser Wavelength Stability," *Instrument Society of America Transactions* 6 (1967), 293–297.

144. "Laser Measures Length to 0.07 ppm," *Instruments and Control Systems* 38 (July 1965), 134–135.

145. A. D. White, "Frequency Stabilization of Gas Lasers," *IEEE Journal of Quantum Electronics* QE-1 (1965), 349–357.

146. The Lamb dip thus presents a nice illustration of the way in which theory and applications interact. For the discovery of the dip and its subsequent elucidation, see W. E. Lamb jr., "Laser Theory and Doppler Effects," *IEEE Journal of Quantum*

Electronics QE-20 (1984), 551–555. See also Nancy Cartwright, "Causation in Physics: Causal Processes and Mathematical Derivations," in P. D. Asquith and P. Kitcher, eds., *PSA 2* (East Lansing, Michigan: Philosophy of Science Association, 1985), pp. 391–404, and N. Cartwright, *Nature's Capacities and Their Measurement* (New York: Oxford University Press, 1989).

147. White, "Frequency Stabilization," cited in note 145. Other methods of frequency stabilization, which was also important for applications such as communications and spectroscopy, are discussed there and also in George Birnbaum, "Frequency Stabilization of Gas Lasers," *Proc. IEEE* 55 (1967), 1015–1026, an article that makes use of White and also of later developments.

148. Mielenz, "Length Measurement," cited in note 143, p. 297.

149. Donald R. Herriott, "Some Applications of Lasers to Interferometry," in E. Wolf, ed., *Progress in Optics VI* (Amsterdam and New York: North-Holland and Wiley, 1967), pp. 171–209.

150. "Laser Interferometer for In-Shop Precision Calibration," *Machinery* 71 (November 1964), 138; W. E. Bushor and J. F. Kreidl, "Where Do We Stand on Laser Length Standards? Part 2," *Laser Focus* 2 (November 1966), 26–29.

151. For Perkin-Elmer and Bendix, see Sheldon Minkowitz, "Industrial Laser Interferometry," *Optical Spectra* (May/June 1968), 64–68, and Winston E. Kock, "The Use of Lasers for Measuring," *Optical Spectra* (March 1970), 60–64, respectively. For Oak Ridge, see Robert W. Schede, "Interferometers for Use as Integral Parts of Machine Tools," *IEEE Transactions on Industry and General Applications* IGA-3 (July/August 1967), 328–332.

152. *Laser Focus* 7 (January 1971), 29.

153. The sales figures are inferred from the annual surveys in *Laser Focus*, January 1970 through January 1973. The number of units sold is inferred from Bushor and Kreidl, "Where Do We Stand?," cited in note 150.

154. See, as examples of U.S. work, V. Vali et al., "Observation of Earth Tides Using a Laser Interferometer," *Journal of Applied Physics* 37 (1966), 580–582, and Jon Berger and R. H. Lovberg, "Earth Strain Measurements with a Laser Interferometer," *Science* 170 (16 October 1970), 296–303.

155. Burns, interview, cited in note 9.

156. Arthur Anderson, paraphrase of an interview, 10 December 1987 (SHL).

157. *Hughesnews* (27 March 1964), Hughes Aircraft Company Archives; see also chapter 4.

158. *Electronics* 37 (19 October 1964), 96.

159. F. P. Gagliano, R. M. Lumley, and L. S. Watkins, "Lasers in Industry," *Proc. IEEE* 57 (February 1969), 114–147.

160. *Electronics* 36 (25 October 1963), 88ff.; *Hughesnews* (27 March 1964), Hughes Aircraft Company Archives.

161. *Raytheon News* (in-house publication) 15 (January 1966), 2.

162. Sidney S. Charschan, interview, 13 November 1984 (SHL); Sidney S. Charschan, "The Evolution of Laser Machining and Welding, with Safety," *Proc. SPIE* 229 (1980), 144–153.

163. J. F. Smith, "Drilling, Trimming, Welding," *Laser Focus* 5 (March 1969), 32ff.

164. *Laser Focus* 4 (August 1968), 12.

165. J. F. Ready et al., "Applications of Effects of High-Power Laser Radiation," *Laser Focus* 2 (15 February 1966), 3–7.

166. Robert D. Haun jr. et al, "The Application of Lasers to Industry," *IEEE Transactions on Industry and General Applications* IGA-4 (July/August 1968), 379–390.

167. "Cutting Cloth by Laser," *Time* 97 (22 March 1970), 71. The poor absorptivity of metals was also overcome by brute-force methods of overwhelming the workpiece with power.

168. A. B. J. Sullivan and P. T. Houdcroft, "Gas-Jet Laser Cutting," *British Welding Journal* 14 (August 1967), 443–445; *Iron Age* (23 April 1970), 99ff. (courtesy of D. H. Belforte).

169. "Laser Market Review," *Laser Focus/Electro-Optics* 23 (January 1987), 53–54.

170. Barry Miller, "U.S. Plan to Accelerate Laser Development Spurs Market," *Aviation Week & Space Technology* 87 (21 August 1967), 92ff., and "Devices Gain Greater Weapons Role," ibid. 92 (19 January 1970), 54ff. Robert W. Seidel, "How the Military Responded to the Laser," *Physics Today* 41 (October 1988), 15; Peter deLeon, "The Laser-Guided Bomb: Case History of a Development," Rand Corporation Report R-1312-1-PR, June 1974.

171. M. I. Cohen, private communication, 21 July 1988.

172. Anderson, interview, cited in note 156.

173. Charles Berry, Oscar Hauptman, and Ronald Kerl, "Origins and Applications of Laser Technology . . .," Term Project Paper for Course 15.367, Sloan School of Management, MIT (courtesy of O. Hauptman).

174. Stanley L. Ream, "Using Lasers for Materials Processing in Industry," Battelle Inputs to Technical Planning, Report #23, 1980, p. 3.

175. *Electronics Industry Association Yearbook*, 1970, p. 63. It is worth remarking that the military use of lasers became a major stimulus to the tasks of defining safe levels of laser power and developing goggles and other protective gear. To protect their personnel from injury and themselves from lawsuits, the services financed research into the biological effects of lasers and into safety standards.

176. S. P. S. Porto and D. L. Wood, "Ruby Optical Maser as a Raman Source," *Journal of the Optical Society of America* 52 (1962), 251–252.

177. J. P. Cedarholm, G. F. Bland, B. L. Havens, and C. H. Townes, "New Experimental Test of Special Relativity," *Physical Review Letters* 1 (1958), 342–344; J. P. Cedarholm and C. H. Townes, "A New Experimental Test of Special Relativity," *Nature* 184 (31 October 1959), 1350–1351.

178. The frequency depends, to first approximation, on the frequencies at which the Fabry-Perot cavity resonates, and these are given by $f = nc/2L$, where c is the velocity of light, L is the length of the cavity, and n is an integer; see T. S. Jaseja, A. Javan, J. Murray, and C. H. Townes, "Test of Special Relativity or of the Isotropy of Space by Use of Infrared Masers," *Physical Review* 133A (1964), 1221–1225. Note that, more strictly speaking, the experiment tests *either* the constancy of light *or* the isotropy of space near the earth.

179. There are some enticing questions that cannot be explored here, because my examples are arbitrarily chosen rather than systematic. What are the differences between the way in which laser physicists fashioned the laser into instruments and the way in which they were used by scientists outside the laser field? How was the spread of laser instrumentation related to the sale of commercial lasers and, toward the end of the 1960s, to the commercialization of instruments such as laser Raman spectrometers? How were applications in science related to technological applications? For a study that provides a pattern for approaching some of these questions, see Yakov Rabkin, "Technological Innovation in Science: The Adoption of Infrared Spectroscopy by Chemists," *Isis* 78 (1987), 31–54.

180. H. Z. Cummins, N. Knable, and Y. Yeh, "Observation of Diffusion Broadening of Rayleigh Scattered Light," *Physical Review Letters* 12 (1964), 150–153.

181. Pierre Bergé et al., "Mise en évidence du mouvement propre de microorganismes vivants. . . ," *Comptes Rendus de l'Académie des Sciences, Paris* 265D (1967), 889–892; R. Nossal, S.-H. Chen, and C.-C. Lai, "Use of Laser Scattering for Quantitative Determinations of Bacterial Motility," *Optics Communications* 4 (1971), 35–39; R. Nossal and S.-H. Chen, "Laser Measurements of Chemotactic Response of Bacteria," *Optics Communications* 5 (1972), 117–121.

182. R. R. Alfano and S. L. Shapiro, "Ultrafast Phenomena in Liquids and Solids," *Scientific American* 228 (June 1973), 42ff.

183. A. L. Schawlow, "Spectroscopy with Tunable Lasers in the Visible Region," in Feld, Javan, and Kurnit, eds., *Fundamental and Applied Laser Physics*, cited in note 95, pp. 667–687. Schawlow shared the Nobel Prize in Physics in 1981 for his work on laser spectroscopy. The other two recipients were Nicolaas Bloembergen, for nonlinear optics, and Kai Siegbahn.

184. Hansch, interview, cited in note 96; T. W. Hansch, "Saturation Spectroscopy of Atoms," in R. G. Brewer and A. Mooradian, eds., *Proceedings of the International Conference on Laser Spectroscopy, Vail, Colorado, 1973* (New York: Plenum, 1974), pp. 353–360; T. W. Hansch, "Repetitively Pulsed Tunable Dye Laser for High Resolution Spectroscopy," *Applied Optics* 11 (1972), 895–898.

185. *Laser Focus* 6 (January 1970), 24; 8 (January 1972), 24.

186. Burns, interview, cited in note 9; for the fall in university R&D resources, see the articles on the subject in the 1969 issues of *Science.*

187. National Science Foundation, "National Patterns," cited in note 8, p. vii.

188. A. J. DeMaria, interview, 13 April 1984 (edited version, 26 December 1985) (SHL).

189. Alexander J. Glass, interview, 13 October 1986 (SHL).

190. Electronics Industries Association, *Electronic Market Data Book, 1971*, p. 67; Neil A. Martin, "On the Beam," *Barron's* (2 February 1970), 11ff.

191. *Laser Focus* 6 (January 1970), 26.

192. Arthur R. Kantrowitz, interview, 30 October 1984 (SHL).

193. *Moody's Industrial Manual*, "AVCO Manufacturing Corporation" and "AVCO Corporation," 1956, 1957, 1960, and 1969. Charles D. Orth III, Joseph C. Bailey, and Francis W. Wolek, *Administering Research and Development: The Behavior of Sci-*

entists and Engineers in Organizations (Holmwood, Illinois: Dorsey-Irwin Press, 1964), p. 550.

194. Seidel, "From Glow to Flow," cited in note 1; John D. Anderson jr., *Gasdynamic Lasers: An Introduction* (New York: Academic Press, 1976).

195. See the preceding references and also Edward T. Gerry, "Gasdynamic Lasers," *IEEE Spectrum* (November 1970), 51–58.

196. The gas dynamic laser's day in the sun of beam weaponry lasted until the early 1970s, when it began to be replaced first by chemical lasers and then by excimer and free-electron lasers. See the epilogue.

197. Kantrowitz, interview, cited in note 192.

198. See "Isotope Separation Process," U.S. Patent 3,443,087. This section draws on a discussion paper written in collaboration with Lee Grodzins for the AAAS Project on Secrecy and Openness in Scientific and Technical Communication and presented as Project Paper No. 8, December 1984.

199. W. B. Tiffany, H. W. Moos, and A. L. Schawlow, "Selective Laser Photocatalysis of Bromine Reactions" *Science* 157 (7 July 1967), 40–43.

200. I. Nebenzahl to J. L. Bromberg, 4 March and 12 August 1983, with enclosures including Nebenzahl, "New Processes for the Separation of Isotopes," 9 April 1969, and "Patent Application," 26 October 1969 (SHL); S. E. Harris, "Selective Photo-Dissociation and Isotopic Separation of Simple Molecules," Proposal to the AEC and the NSF, December 1970 (SHL); V. S. Letokhov, "Laser Isotope Separation," *Nature* 277 (1979), 605–610.

201. Irving Itzkan and Frederick W. Cunningham, "Oscillator-Amplifier Dye-Laser System using N_2 Laser Pumping," *IEEE Journal of Quantum Electronics* QE-8 (1972), 101–105; James A. Myer et al., "Dye Laser Stimulation with a Pulsed N_2 Laser Line at 3371A," *Applied Physics Letters* 16 (1970), 3–5. The consultant was MIT professor Lee Grodzins.

202. Testimony of Raymond L. Dickeman, pp. 178–195, "Future Structure of the Uranium Enrichment Industry," JCAE Hearings, Part II, October 1973 (93rd Congress, First Session).

203. Kantrowitz, interview, cited in note 192; G. Sargent Janes, paraphrase of an interview for the Laser History Project, 19 September 1984 (SHL).

204. Other fuels with lesser environmental effects have also been studied, including pure deuterium and deuterium and light helium.

205. The reaction is D + T = He + n, and the energy that is released is in the form of the kinetic energy of the helium and neutron reaction products.

206. Joan Lisa Bromberg, *Fusion: Science, Politics, and the Invention of a New Energy Source* (Cambridge: MIT Press, 1982).

207. F. J. McClung and R. W. Hellwarth, "Giant Optical Pulsations from Ruby," *Journal of Applied Physics* 33 (1962), 828–829. Their peak power was 600 watts, and this is the basis for my energy estimate. R. E. Kidder, "A Brief Account of the Early History and Goals of the LRL Laser Research Program (1961–1963)," 10 August 1972, 3 pp. (courtesy of R. W. Hellwarth). Ray E. Kidder, interview, 20 August 1986 (SHL); T. C. Merkle (Associate Director, Lawrence Radiation Laboratory) to Maj.

Gen. A. W. Betts (Director, Division of Military Application), 23 October 1963 (courtesy of R. E. Kidder).

208. See the references cited in the previous note and John S. Foster jr. to Keith A. Brueckner, 22 May 1962 (courtesy of R. E. Kidder).

209. J. E. Swain, R. E. Kidder, K. Pettipiece, F. Rainer, D. Baird, and B. Loth, "Large-Aperture Glass Disk Laser System," *Journal of Applied Physics* 40 (1969), 3973–3977.

210. N. G. Basov et al., "Experiments on the Observation of Neutron Emission at a Focus of High-Power Laser Radiation on a Lithium Deuteride Surface," *IEEE Journal of Quantum Electronics* QE-4 (November 1968), 864–867. Gloria B. Lubkin, in "Search and Discovery," *Physics Today* 21 (November 1968), 57, and 22 (November 1969), 55.

211. KMS Fusion, Inc., "Partial History of Laser Fusion As It Pertains to KMSF," 10 November 1971 (courtesy of Roy R. Johnson) (SHL).

212. Bromberg, *Fusion*, cited in note 206, pp. 175–187.

213. Ignition is the point at which the temperature of the fuel is maintained by the energy of the fusion reaction products, so that no further outside energy is required.

214. J. Nuckolls, L. Wood, A. Theissen, and G. Zimmerman, "Laser Compression of Matter to Super-High Densities: Thermonuclear (CTR) Applications," *Nature* 239 (15 September 1972), 139–142.

215. These problems in turn created new demands on the lasers, since the way in which the laser beam interacts with the pellet is affected by the laser's wavelength. By 1980, this had led to the decision that it was necessary to use small wavelengths, and had also led to the realization that breakeven would require energies closer to 1 million joules than to the thousand joules that Nuckolls and Wood had originally supposed. See the epilogue to this book and also William D. Metz, "Ambitious Energy Project Loses Luster," *Science* 212 (1 May 1981), 517–519.

216. Keith A. Brueckner, autobiographical memoir, 1986, Niels Bohr Library, American Institute of Physics. KMS "Partial History," cited in note 211.

217. Gene Bylinsky, "KMS Industries Bets Its Life on Laser Fusion," *Fortune* 90 (December 1974), 148ff.

218. M. J. Lubin, private communication, 25 September 1984; Lubkin, in *Physics Today* (November 1969), cited in note 210; Robert L. Hirsch, "Laser Produced Plasmas for Controlled Thermonuclear Research: Perspective 1969," September 1969, U.S. Atomic Energy Commission.

219. Keith Boyer, interview by Robert W. Seidel, 5 November 1984 (SHL); Kidder, interview, cited in note 207.

220. Lubkin, in *Physics Today* (November 1969), cited in note 210.

221. Kidder, interview, cited in note 207.

222. "Laser Fusion: Pipeline to the Sun," promotional pamphlet (courtesy of the University of Rochester Laboratory for Laser Energetics).

223. This comprised operating funds, construction, and capital equipment. Five years later, in fiscal 1980, the figure stood at $195 million. (Figures in current dollars; source: U.S. Department of Energy.)

Chapter 6: Explaining the Laser

1. See, for example, C. H. Townes, "Ideas and Stumbling Blocks in Quantum Electronics," *IEEE Journal of Quantum Electronics* QE-20 (1984), 547–550; Irina Mikhailovna Dunskaya, *Vozniknovenia Kvantovoi Elektroniki* [Origins of quantum electronics] (Moscow: Isdatel'stvo Nauka, 1974); partial English translation by Patricia Sobchak, typescript, courtesy of A. L. Schawlow and C. H. Townes. Michael S. Feld, "Historical Development of the Maser," Part I of "Absorption Spectroscopy Using Optical Masers," BS/MS thesis, Department of Physics, MIT, 1963; Mario Bertolotti, *Masers and Lasers: An Historical Approach* (Bristol: Adam Hilger, 1983), pp. 7, 11, 27, 32.

2. Dunskaya, book cited in note 1, chapter II. Bela A. Lengyel, "Evolution of Masers and Lasers," *American Journal of Physics* 34 (1966), 903–913. F. G. Houtermans, "Uber Maser-wirkung im optischen Spektralgebeit und die Möglichkeit absolut negativen Absorption für eininge Fälle von Molekülspectren (Licht-Lawine)," *Helvetica Physica Acta* 33 (1960), 933–940. These sources give no information on the context of ideas within which these men were working, information that is, of course, indispensable for judging the relation of their theories to lasers. Fabrikant's 1950s work on light amplification does mean that in the Soviet Union, work on lasers *preceded* work on masers. This does not vitiate my analysis, which is only framed for U.S. conditions. It does, however, point up the interest of having histories of Soviet quantum electronics to compare with U.S. developments.

3. Dunskaya, in the book cited in note 1, cites their paper, "The Negative Absorption of Radiation," *Nature* 122 (1928), 12–13.

4. Bertolotti, *Masers and Lasers*, cited in note 1, chapter 3, pp. 32–72. See also Nicolaas Bloembergen, interview, 27 June 1983 (SHL).

5. Feld, "Historical Development," cited in note 1.

6. Townes, "Ideas and Stumbling Blocks," cited in note 1, p. 547.

7. C. H. Townes, "Microwave Spectroscopy," *American Scientist* 40 (1952), 270–290.

8. This line of thought immediately suggests comparing "hot topics" at different points in time. A first profitable comparison might be of the laser boom of the early 1960s with the high-temperature-superconductor boom that commenced at the end of 1986.

9. H. G. J. Aitken, *The Continuous Wave: Technology and American Radio, 1900–1932* (Princeton: Princeton University Press, 1985).

10. A. Hunter Dupree sounded an early call for this analysis in "Influence of the Past: An Interpretation of Recent Development in the Context of 200 Years of History," *Annals, American Academy of Political and Social Science* 327 (1960), 19–26, and "The Structure of the Government-University Partnership after World War II," *Bulletin of the History of Medicine* 39 (1965), 245–251.

Index

Abella, Isaac D., 78, 93
Absorption
    in molecular systems, 16–17
    and Q-switching, 186
    and transparency, 101
Academic research
    on lasers, 208
    on masers, 7–8, 30
Active satellites, 56
Adams, C. M., jr., 180
Adler, Richard B., 44
Advanced Research Projects Agency
    (ARPA), 4, 57, 82, 87, 101–102, 158,
    176, 211
Advisory Committee on Millimeter
    Wave Generation, 15–16, 19
Aigrain, Pierre, 143–144, 153
Airborne Instruments Laboratory, 26–
    27
Aitken, Hugh G. J., 75, 226
Akhmanov, S., 185
Alignment applications, 162
Alsop, Leonard E., 49, 52
American Optical Company, 5, 7, 10,
    90, 99, 101, 130–131
American Physical Society, 103
Ammonia absorption clock, 22
Ammonia beam maser, 13–31
    and solid-state masers, 31–46
Ammonia molecules, 21
ANTARES, 241
Applications, 12, 190–208
    alignment, 162
    of argon ion lasers, 171
    future of, 248
    industrial, 231–232
    for length measurement, 190, 197–
    200
    medical, 211, 243–246

for particle velocity measurement,
    207
photocoagulation, 130–131, 159,
    161, 171, 243–245
    See also Military
Applied Optics, 104, 111, 113
Applied Physics Letters, 151
Argon as buffer, 167
Argon ion laser, 161, 165, 167–168
    applications of, 171
    for coagulation, 243
    continuous, 168–169
    efficiency of, 170
    for glaucoma, 245
    lifetime of, 171, 173
    for materials processing, 198, 203
    power from, 169–170
    problems with, 171–173
Argon-oxygen molecular-dissociation
    lasers, 141
Arsenic ions in ammonia masers, 32–
    33
Artman, Joseph O., 38, 41, 48
Astronomy, 46, 48–49, 51, 58–59, 105
AT&T, 55
    patents of, 124–125
    See also Bell Telephone Laboratories
Atmospheric effects on laser commu-
    nications, 191, 193–196
Atomic clocks, 24
Atwood, John, 120
Auclair, Jean-Michel, 211, 212
Audio discs, 232
AVCO Everett Research Laboratories
    (AERL), 209–213, 241, 243
Aviation, masers for, 46
Aviation Week and Space Technology, 99,
    102

Background radiation, 57, 59–60

Printed in the United States
by Baker & Taylor Publisher Services

Printed in the United States
by Baker & Taylor Publisher Services